Ancient Views on the Origins of Life

Ancient Views on the Origins of Life

Ernest L. Abel

Rutherford • Madison • Teaneck
Fairleigh Dickinson University Press

© 1973 by Associated University Presses, Inc.

Associated University Presses, Inc.
Cranbury, New Jersey 08512

Library of Congress Cataloging in Publication Data
Abel, Ernest L   1943–
  Ancient views on the origins of life.

  Bibliography: p.
  1. Life—Origin.   I. Title.
QH325.A23   577   72–656
ISBN 0-8386-1198-2

Printed in the United States of America

To Barbara

Contents

Preface		9
Acknowledgments		11
1	The Origins of Life in Mythology	15
2	The Emergence of the Scientific Spirit	22
3	Pre-Socratic Scientific Accomplishment	50
4	War and the Demise of Greek Science	54
5	Aristotle's Views on the Origins of Species	59
6	Theophrastus's Doubts Concerning the Theory of Spontaneous Generation	64
7	The Quest for Self-Awareness	66
8	Early Christian Reappraisals	70
9	Factors in the Decline of Science	75
Epilogue		77
Notes		81
Bibliography		87
Index		91

Preface

My purpose in writing this short book is to provide those who are interested in the history of ideas with an account of how the people of the ancient world tried to explain the appearance of man and other forms of life on earth.

In setting about this task, my first impulse was to revise and update H. F. Osborn's *From the Greeks to Darwin,* which appeared in 1894. This book still remains the main source from which many historians of biology draw their information in discussing the way in which the evolution-idea was thought about in Antiquity. But it was soon apparent from rereading Osborn's work that more than a simple revision would be required. For one thing, numerous articles on various aspects of this subject have since appeared in sundry scientific and philosophical journals. For another, Osborn had completely ignored the intellectual climate within which the Ancients interpreted their observations. In light of the many new studies of the original source material, an updated presentation of the subject seemed necessary.

What follows, then, is a new outline, encompassing as far as possible not only the history of the idea, but also the social and philosophical factors that influenced the course of its development.

Acknowledgments

The author is grateful to the following publishers for permission to quote from copyrighted material:

Basil Blackwell Publishers, for permission to quote from K. Freeman, *Ancilla to the Pre-Socratic Philosophers*, 1948.

E. J. Brill, Leiden, for permission to quote from J. H. Loenen, "Was Anaximander an Evolutionist?" in *Mnemosyne* 7 (1954): 215–32.

Cambridge University Press, for permission to quote from G. S. Kirk and J. E. Raven, *Presocratic Philosophers*, 1957.

The Clarendon Press, for permission to quote from Aristotle, *The Oxford Translation of Aristotle*, ed. W. D. Ross, 1928. By permission of the Clarendon Press, Oxford.

Harvard University Press, for permission to quote from T. S. Hall, *A Source Book in Animal Biology*, 1964 (Francesco Redi's *Experiments on the Generation of Insects*, translated by M. Bigelow, Chicago, 1909) and the Loeb Classical Library editions of St. Augustine (*City of God*), Strabo, Diodorus Siculus, Hesiod (*Works and Days*), Dio Chrysostom, and from K. Freeman, *Ancilla to the Pre-Socratic Philosophers*, 1948.

Hodder and Stoughton Limited, for permission to quote from

S. G. F. Brandon, *Creation Legends of the Ancient Near East*, 1963.

Random House, for permission to quote from W. J. Oates, *The Stoic and Epicurean Philosophers*, 1940.

Routledge & Kegan Paul Ltd., for permission to quote from S. Sambursky, *The Physical World of the Greeks*, 1960.

Ancient Views
on the Origins of Life

1
The Origins of Life in Mythology

The theory of evolution holds that all organisms are the result of descent and modification from a simple, undifferentiated primordial substrata. Although it is customary to credit the inception of this theory to Charles Darwin and his immediate predecessors, a rudimentary form of this notion can be traced back to the beginnings of written history itself. In fact, the belief that life had its origins in a single basic substance is so widespread among the various peoples of the world, primitive or civilized, that it can be considered one of the few universal themes in the history of ideas.

Two consecutive phases in the development of this conception can be distinguished. The first involves the mythical interpretation of natural phenomena; the second, the rational or philosophical. This dichotomy, however, is made only by our modern-day historians. The ancient chronicler perceived no such finite classification. "Even a lover of myth," wrote Aristotle, "is in a sense a philosopher."[1]

In general, myth conveys the impression of a story invented *ex nihilo*, a story describing the irascible and typically irresponsible actions of various divine malcontents. But these deities are not simply malevolent gods capriciously toying with mankind. They are actually personifications of Nature, and their activities, predictable and unpredictable, determine what life will be like on earth. For example, the Greek god Apollo, or his Egyptian counterpart, Amon Re, are at once the sun and the unfathomable divinities who cause the earth to be lighted by driving their blazing chariot across the skies. It may have been naïve to attribute the sun's movements to unseen powers, but the superstition did not prevent men from making accurate and precise records of these divine activities. And after all, precise observation is the beginning of science. The actual distinction between the mythical and scientific approach to Nature lies not in the observations upon which they are based, but in the interpretation of these observations.[2]

It has already been noted that the concept of life emerging from a single substrate is a notion held by many of the primitive peoples of the world. As such, it was also one of the most recurrent themes in the myths of antiquity. Every nation and city-state had such a myth, and every society claimed that its ancestors were the first humans to have appeared in the world. Even the Greeks believed that they were autochthonous, even though they were fully aware that the civilizations in Egypt and Mesopotamia predated their own by more than a thousand years.

But even though this belief is universal in its distribution, the observations that prompted the idea remain a mystery for all but a few civilizations. Fortunately, Egypt is one of these exceptions. In the first century B.C. the Greek historian, Diodorus of Sicily, visited the country and he recorded the following observations: "even at the present day the soil of Thebes at certain times generates mice in such size as to astonish all who have witnessed the phenomenon; for some of them are fully formed as far as the breast and front feet, and are able to move, while the rest of the body is unformed, the clod of earth still retaining its natural character."

These observations prompted Diodorus to conclude: "for this fact it is manifest that, when the world was taking shape, the land of Egypt could better than any other have been the place where mankind came into being because of the well-tempered nature of its soil: for even at the present time, when the soil of no other country generates any such things, in it alone certain living creatures may be seen coming into being in a marvellous fashion.... Indeed, even in our own day during the inundations of Egypt the generation of forms of animal life can clearly be seen taking place in the pools which remain the longest; for, whenever the river has begun to recede and the sun has thoroughly dried the surface of the slime, living animals, they say, take shape, some of them fully formed, but some only half so, and still actually united with the earth."[3]

Thus, the conviction that life had initially emerged from the soil was based on certain evidence to this effect that was still observable in the first century B.C. When the Egyptians saw what appeared to be the autochthonous emergence of animals from the earth, it is not surprising that they concluded that their own ancestors had also emerged from the soil.

The earliest testimony to this belief comes from an Egyptian bas-relief dating from about 1400 B.C., which depicts the god Khnum sitting at a potter's wheel upon which stand two children.[4] The idea is that man was originally fashioned out of clay—an idea that in the seventh to sixth century B.C. was expressed in writing with the declaration that "'man is clay and straw and god is his builder."[5] The god Khnum had modeled man's body upon a potter's wheel and had placed the prototype in the soil. Then, when it was time to appear, man simply emerged from the earth just as did the other creatures.

In Babylonia, man was also thought to have been fashioned out of clay, but coupled with this belief was the impression that the gods had had a special purpose in creating man, namely, to provide themselves with servants. For example, in one myth the Mother Goddess, Lullu, is ordered to create a being to serve all of the gods, and to build him out of clay.[6] Similarly, in the creation myth called the *Enuma elish,* the god Marduk declares that he will bring together blood and bones and from these he will create a "savage," which shall be named *man.* This savage will be burdened with serving the gods so that the deities will be able to live leisurely existences.[7]

The explanation for this self-denigrating attitude is readily apparent when one recognizes that life in Mesopotamia involved a yearly struggle for survival against the elements. Unlike the Nile, the Tigris and Euphrates rivers were not predictable in their actions. Without warning they might crash down from their sources in the Armenian mountains. The Mesopotamian plains would be suddenly inundated, the crops destroyed before they could be harvested. The inhabitants lived at the very mercy of the rivers, despite the elaborate network of canals that they built to check and distribute these waters. The uncertainty of day-to-day life engendered a sense of despair and fatalism. It is no wonder then, that these people saw themselves in the role of the oppressed slaves.

By contrast, Egypt was a country that was relatively free from natural catastrophes. The Nile flooded the land with dependable regularity and men had only to plant their seeds to be assured of food. This sense of security was reflected in a feeling of confidence and optimism on the part of the Egyptians, and it enabled them to devote more of their time to further technological mastery over their environment.

The remaining myth of creation indigenous to the Near East is found in the Old Testament. Actually, there are two independent narratives contained in these writings; the older of the two myths was written some time between 900 and 500

B.C. and begins at Genesis 2:4b, while the second, written about 500 B.C., runs from Genesis 1:1–2:4a.

The older story is called the Yahwist account, because in it God is addressed as Yahweh. The beginnings of the universe are completely ignored by the chronicler. Instead, the story commences with Yahweh fashioning "man of dust from the soil."[8] "Yahweh (then) said, 'It is not good that man should be alone. I will make him a helpmate.' So from the soil God fashioned all the wild beasts and all the birds of heaven."[9]

Since the origins of the Hebrews are rooted in both Egypt and Mesopotamia, it is a moot point whether the writer borrowed or conceived on his own the idea that God had fashioned man out of clay. Such myths can be found throughout the inhabited world[10] so there is no reason to assume that the Yahwist historian was simply repeating a foreign myth of the creation. Whatever the source, it is to be noted that man and the animals share a common beginning in this particular version, each having been brought to life from dust by Yahweh's own hand. Also worth noting is the sequence of appearance: man first, then the animals.

While the second account, called the Priestly source, is essentially a commentary on the Yahwist story, it does introduce a few rather significant additions and changes. For one thing, it begins with the formation of the universe and it implies that there was a preexistent watery state from which land was made to appear. From this land arose vegetation and "every kind of living creature." As his final act, God creates man "in the image of himself."

Two changes have thus been made in the original narrative. Man is now depicted as having been created after, rather than prior to the animals, suggesting perhaps that man represents the height of God's magnificent accomplishments. Second, while the other forms of life are commanded to appear by fiat, the Priestly source states that man was personally fashioned by God's own hand.

Although often venerated for their rationality, during their early history the Greeks also held a credulous belief in the existence of inscrutable and progenitive gods. According to the poet Hesiod,[11] who lived around 800 B.C., Hephaestus had been commissioned by Zeus to "mix earth with water and to put in it the voice and strength of human kind." Four successive but discontinuous creations of this sort were envisioned. After the first, called the "golden race of mortal men," there appeared the silver and then the bronze—the men who settled in Greece around 2000 B.C. After these came the race of "half-gods," who raided Troy, and then finally the race of men to which Hesiod himself belonged—the race of iron.

In a later cycle of myths, the Titan god Prometheus is venerated as the benefactor of mankind for having created man out of earth and water and for presenting him with fire, a gift that he stole from the Olympian gods. Enraged by the theft, Zeus had the Titan chained to a cliff where he was visited daily by a monstrous eagle that tore at his liver. Other Greek legends assert that during an abortive rebellion against Kronos, man arose from the blood of the slain Titans.[12]

In a still later story, Prometheus is said to have been freed and to have had a son named Deucalion who wedded Pyrrha, the daughter of Pandora, the first woman to have been created. The two are warned by Prometheus that Zeus is going to destroy mankind because of its wickedness, and like Noah of the Old Testament, the couple build an ark and ride out the flood that Zeus sends.

When the rains finally cease and the ark comes to rest on dry land, Deucalion disembarks and sacrifices to Zeus. Zeus is appeased and as a token of good will he declares that he will grant Deucalion any wish that he might have. Deucalion's immediate request is for companions. Zeus realizes his mistake but it is too late: his promise has been given. Deucalion and his wife are told to throw the "bones of their mother" over their shoulders and his wish will be fulfilled. At first Pyrrha is disquieted at the order, but Deucalion comforts her with the explanation that Zeus means the "bones of Mother Earth." Each then picks up a number of stones. Those thrown by Deucalion become men, while those thrown by Pyrrha appear as women. Thus, mankind once again is made to rise from Mother Earth and this is the reason that these creatures are called *laos* (people), for they came into being from *laas,* the Greek word for stone.

It should now be apparent that the belief in the autochthonous origins of mankind was not confined to any one area of the ancient world, but was shared by many early civilizations, among them the Egyptians, Babylonians, Hebrews, and Greeks. In the Near East this idea was accepted as incontrovertible, owing to the religious endorsement that it received. In Greece, however, the only religious requirement confronting the Greek was that he recognize the existence of the gods. Following this, he was free to say or write whatever he thought about them or their actions. Given this religious freedom, the Greeks eventually began to ask themselves questions that either did not arise in these other societies, or if they did arise, were quickly suppressed by the powerful priestly classes.

2
The Emergence of the Scientific Spirit

I

Although the Egyptians and Babylonians made great strides in the exact sciences of mathematics and astronomy, they never advanced beyond the stage of collecting and cataloguing individual facts. No generalizations were ever ventured in their scientific writings and no attempt was ever made to postulate basic principles. The facts that were recorded were never applied to new circumstances or conditions. Mathematicians solved difficult problems, but never proposed a theorem or axiom. Historians recorded significant contemporary events, but failed to place them in the perspective of past events.

For the most part, the major impediment to scientific progress in the Near East followed from the assumptions that these people held about their universe. While we today assume that events in Nature can be fully explained as the result of specifiable causal factors, the Egyptians and Babylonians assumed that the course of events in Nature was indeterminate and therefore beyond the understanding of mankind. Consequently, they made no effort to seek lawful relationships between various recurrent events, nor did they ever gain any deep insight into the events that they observed and recorded.

This in turn engendered an attitude that knowledge was important only for the uses to which it might be put. For instance, when the Babylonian priests turned their attention to natural phenomena such as the stars, it was for the purpose of knowing better the will of the gods whose actions were believed to be intimately linked with the movements of these heavenly bodies. The stars themselves were of secondary interest. Likewise, the Egyptians invented geometry so that they could erect huge testimonials to their kings and never did they abstract the theoretical principles that lay behind their calculations.

Tradition credits Thales of Miletus in Ionia with many scientific accomplishments, among them a prediction of the eclipse of the sun that occurred in 585 B.C.[1] and the introduction of geometry into Greece. However, Thales' real contribution to science lies not in these legendary deeds, but in the influence that he exerted upon all subsequent thinking about Nature.

Hitherto, men had resorted to fanciful tales about their gods to account for phenomena such as the clouds and the rains. Thales, by contrast, emphasized Reason as the key to understanding occurrences in Nature, and assumed that by observing these events, he could discover the laws under which Nature operated from day to day.

The actual turning point in the history of science began when Thales declared that the gods played no personal role in Nature. This is not to say that he denied the existence of the gods. Not at all. Thales still accepted Hesiod's postulation of a gradual unfolding in Nature and he still identified the elements with the gods ("all things are full of gods"), but he did challenge the premise that these deities had willful personalities. They were, so to speak, nothing more than automatons. Thus freeing himself from the shackles of mythological indeterminism, Thales confidently began to probe Nature for its hidden secrets.

As his first endeavor, Thales turned his attention to the same mystery that had prompted many of the ancient myths: the material or materials from which the world was initially formed. Accepting Hesiod's basic premise of a progressive differentiation in Nature, Thales reasoned that if there were a chain of events through which the world had passed, then by following that chain link by link, he might be able to determine the *arche,* the primary world-stuff (*Weltstoff*) out of which the world had been formed.

In the end, observation and introspection caused him to conclude that all of the variations in Nature could in fact be accounted for in terms of a single substance—water. According to Aristotle, this opinion followed from Thales' observation "that the nourishment of all things is moist . . . and from the fact that semen is moist, and that all living things depend upon water for their existence."[2] But Aristotle admitted that he himself was only guessing concerning Thales' evidence.

In Plutarch's opinion,[3] Thales' choice of water as the first principle was the result of his travels to Egypt, where he had observed the flooding of the Nile and had seen how the very existence of the country depended upon this inundation. Commerce between the cities of Miletus and Naucratis, the Ionian colony in Egypt, was rather commonplace at that time, and it is probable that a citizen of Thales' prominence would have visited Egypt at least once in his life. Having seen the dependence of that country on the Nile, Thales could not help being impressed by the lifegiving potential of water. But whatever it was that prompted him to choose water as the source of all things, historians both ancient and contemporary agree that Thales' decision was based on observation and reason, and that no appeal was made to the supernatural, as had previously been the custom. And as a result of this approach to Nature, Thales has rightly been hailed as the "first man of science."[4]

Thales, however, failed to explain what it was that caused water to change into other forms of matter. Nor did any of his pupils ever distinguish the animate from the inanimate. It was simply assumed that matter contained its own energy. Perhaps Thales was thinking about this self-contained property of energy within matter when he declared that "all things are full of gods," but clearly this was a question that had to be resolved in order for science to free itself from the notion of a *deus ex machina*.

Thales' chief pupil and successor was Anaximander of Miletus (fl. 565 B.C.), who is credited with constructing the first map intended for navigational use. Like Thales, Anaximander also accepted the doctrine of a single elementary and indestructible substance from which all matter originated, but he disagreed with his teacher as to its identification. Rather than some recognizable substance such as water, Anaximander held that the "first principle" was infinite, imperceptible, and therefore undefinable, and to avoid any inappropriate analogies being applied to it, he called this basic material the *Apeiron,* meaning "boundless." According to Anaximander, the *Apeiron* originally contained all the primary elements of matter: hot (fire) and cold (air) and wet (water) and dry (earth). These eventually separated from the *Apeiron* and through the counter-antagonism and intermingling of these four elements, the material basis of the world came into existence.

Anaximander also appears to have had definite ideas concerning the beginnings of life. In these ruminations (which the eminent historian of philosophy, J. Burnet, calls Anaximander's "crowning audacity"[5]), the Milesian philosopher appears to introduce what some historians[6] have seen as a close approximation to Darwin's concept of the mutability of species.

Others,[7] however, have argued that no such interpretation of Anaximander's theory is warranted. Since Anaximander's own writings are no longer extant, the secondary evidence upon which this dispute is based must be critically examined. It may be that these secondary sources are interpretations rather than faithful quotations or paraphrases of Anaximander's own words, hence it is to this source material that we now turn:[8]

1. The first living beings were generated in the moist element, each enveloped in a prickly bark. Growing older they went on dry land and after the bursting of the bark they changed their way of living (Aetius, 5,19,4).

2. Man originally was born from living beings of another species, because man alone of all living beings needs long care in infancy; therefore he would never have survived, if he always had been as he is now (Ps.—Plutarch, *Strom.*[2]).

3. Animals came into being from the moist element when it was evaporated by the sun. Man, however, came into being from another animal, namely, the fish, for in the beginning he was like a fish (Hippolytus, *Ref.* 1,6,6).

4. Originally men came into being inside fishes and having been fed like sharks they came out as soon as they were able to look after themselves and went ashore (Plutarch, *Symp.* 8, 730E).

5. From water and earth originated first, under the influence of the heat of the sun, fishes or fish-like creatures; inside them men came into being and they remained there till the time they had grown up; then the fishes burst and adult men and women who henceforth could feed themselves made their appearance (Censorinus, *de die nat.* 4,7).

Can these statements be accepted as evidence for a primitive form of Darwin's theory; one that was announced almost 2,000 years before Darwin? The most thorough discussion of this problem and at the same time the most recent, is that by J. H. Loenen.[9]

Loenen argues that the function of the "prickly bark" in Anaximander's description of the first animals, was primarily to protect them from the water in which they had originated. Upon leaving the water, this covering was lost, but this loss, Loenen contends, should not be taken as an indication of an evolution of marine into land animals. Instead, it means that Anaximander believed that these marine and land animals were one and the same; the prickly bark was just a protective covering so that land animals could survive in the water.

Secondly, while Anaximander seems to suggest that once the animals reached land, they "adapted" (i.e., "changed their way of living") to their new conditions of life, he did not imply a biological elimination of all those species that were not fit to exist in the new environment. Rather, he felt that upon encountering their new conditions, animals modified their already existent behavioral patterns so that they could survive in the new settings. Thus, while Anaximander did champion the notion of adaptation, the kind of adaptation he had in mind was behavioral, not organic.

Finally, it is to be noted that Anaximander made no reference to the behavior of mythical deities to account for the origins of life. Furthermore, whenever it was possible, he introduced analogies from Nature to corroborate his views. Fanciful though they were, these ideas still bear the mark of sober reflection on a difficult problem.

For instance, in acknowledging the fact that man cannot fend for himself at birth, Anaximander felt that he had to account for man's early years of existence. To have placed man in a "prickly bark," floating in water, would have been to put him in a situation wherein his biological needs could have been met only with great difficulty. Consequently, he relegated man's early years to an existence within the belly of a fish or fish-like creature. Anaximander's choice of the shark as man's host is explained by Plutarch.[10] Apparently, a common observation of the day was that while the shark laid eggs as did the other fish, unlike them, it hatched them not outside but within its own body, and continued to nurture them from within until the offspring were capable of providing for themselves. By analogy, Anaximander believed that the first humans were cared for in the same manner. When they were eventually capable of independent survival, they were freed from the womb of the shark to be cast out upon the shore.

But this does not mean that Anaximander implied that man and the shark had anything biologically in common. What had happened was that man had spontaneously come into being within the fish, just as the fish had spontaneously appeared in the water. Then when they were ready to leave their fish wombs, the first humans emerged fully formed and ready to begin life on land. Had man been able to care for himself at an early age, Anaximander undoubtedly would have surrounded him with a "prickly bark" like that covering the other land animals that arose from the sea.

In summary, Anaximander is not to be thought of as the first advocate of the mutability of species. Rather, his ideas are to be viewed as an early rational attempt to prove the legitimacy of the Abiogenesis hypothesis that life had emerged spontaneously from the sea.

Anaximander's theories, however, were soon challenged by the last of the important Milesian philosophers—his own pupil, Anaximenes (fl. 545 B.C.). Although he agreed with his teacher that the primordial substance was infinite, Anaximenes could not accept Anaximander's indeterminate concept of the *Apeiron* as being the "first principle." Instead, he urged that air was the material substance from which all things had come into being. Air, he stated, was transmutable and in constant motion. When it rarefied, it became fire; when it condensed, it became first "wind, then cloud, then (when thickened still more) water, then earth, then stones, and the rest came into being from these."[11] In essence, Anaximenes had conceptually reduced all matter to quantitative transformations of air, with living creatures consisting of nothing more than "homogenous air and wind."[12]

According to Plutarch,[13] the basis for this conclusion was Anaximenes' observation that air felt either hot or cold depending upon whether one was breathing in or out. When it was compressed and condensed by the lips during inhalation, it felt cool; when it escaped from the mouth during exhalation, it rarefied and therefore felt warm. Since both hot and cold are qualities fundamental in matter, and since they both owed their existence to air, Anaximenes reasoned that matter must also owe its existence to this basic substance.[14]

This theory of the origin of matter may seem less inspired than that of Anaximander's, but, in fact, it was Anaximenes' conception of matter that stirred the imaginations of his contemporaries. It stated plainly of what the "first principle" consisted, but more important, it accounted for every other form of matter in quantitative terms. As such, it came to epitomize Milesian cosmological thinking. Thales had stated that the origin of all things was water; Anaximander had argued that it was impossible to determine the form of the original substance and had called it the *Apeiron* to avoid analogies; but neither of them had entertained the possibility of a relationship between matter and the forms of life that existed on the earth. Anaximenes, on the other hand, explained that there was an element common to both, and any differences between them were due solely to the place that an object occupied in the continuum of air, the basic substance. Thus, it is to Anaximenes, rather than Thales or Anaximander, that we owe the idea that there is a continuum in life, an idea that is fundamental to the modern theory of evolution.

II

In 494 B.C. the Persians overran Miletus, thereby putting an end to its spirit of inquiry as well as its independence. The Ionian city of Ephesus became the new seat of Greek philosophy, with Heraclitus (fl. 500 B.C.) its recognized spokesman.

According to Diogenes Laertius,[1] Heraclitus was a misanthrope and a hermit who shunned the company of others and lived on grasses and plants. Before he became a recluse, however, he is said to have written a book entitled *On Nature*, in which he expressed a number of opinions concerning mankind and Nature. The fragments of this work that are still extant indicate that these opinions were generally unconnected and seem to have been written as the thought struck him. Yet they were so laconic and intriguing that his contemporaries were immediately fascinated by his remarks. Consequently, even though Heraclitus himself had no disciples, his ideas continued to be reflected upon and taught by succeeding generations of philosophers. In the present study only Heraclitus's pronouncements concerning the concept of change and his position regarding the character of the primary substance will be discussed.

Heraclitus essentially agreed with Anaximenes' thesis that there was a continuum in Nature, but he carried the proposal one step further in declaring that changes were constantly taking place in the composition of matter. "One could not step twice in the same river"[2] he argued, for "it scatters and gathers . . . it comes together and flows away . . . approaches and departs."[3]

The Milesian philosophers had also recognized that change was a concept that was fundamental to the monistic theory of a single substance giving rise to other forms of matter. But none of them had dealt with the source of this fundamental property. Heraclitus's solution was that change was an innate characteristic of matter and therefore it was inevitable that one form of matter give rise to another.

The "first principle," however, was neither water, the *Apeiron,* or air, as the Milesian philosophers had proposed, for none of these materials embodied the notion of change. To Heraclitus, the only possible substance that could qualify was fire.

To us, this choice seems difficult to understand, since fire is a chemical process, not an element. To the ancients, however, fire was considered a basic substance, as were air and water.[4] And in fire, Heraclitus thought he saw the quintessence of flux. Fire was the one substance that was constantly changing in appearance. The changes could even be seen with one's own eyes, and in Heraclitus's own words, "the things of which there is seeing and hearing and perception, these do I prefer."[5]

Thus, in the world order of things, matter consisted of varying amounts of fire. Earth and water were fundamentally fire that had been extinguished in some measure. Presumably, Heraclitus would have proposed the same kind of explanation to account for the differences between living creatures, but if so, there is no record of such conjectures in his extant writings.

Heraclitus's main influence on his successors, however, was not the result of his postulation of fire as the primordial element. Instead, it was his emphasis on change as the basic characteristic of all things that impressed subsequent philosophers. In fact, it was because this postulate lent support to Anaximenes' doctrine of the continuum in Nature that Heraclitus was venerated as one of the leading contributors to pre-Socratic science. But even more significant was the importance he attached to phenomena that were observable as the kind of evidence upon which reasoning was to be based. It is this approach which set Heraclitus off as one of the true pioneers of the scientific method.

III

During this formative period of Ionian philosophy, the trend in thinking was moving more and more away from the mythical and toward the rational examination of Nature. But the period was also one that was characterized by disagreement. Men were asking profound questions, but no one was able to arrive at a satisfactory conclusion. This was because the philosophers of the period rarely ventured beyond guessing at the common properties in the phenomena they observed. Rarely did they actively seek to corroborate their hypotheses with objective evidence. The collection of data to support a particular position was rarely appreciated. The philosophers did not deny the value of data; it was just that they were too impatient to bother with what they probably considered to be minor details.

Consequently, in such matters as the essence of the "primary substance," each of the Ionians based his decision upon a few gross observations, intuition, and what seemed to be common sense. But what was common sense to one was an error in judgment to another. For every common-sense argument that "absence makes the heart grow fonder," someone else might come along and counter that "out of sight, out of mind." What was sorely needed for any further scientific development was objective and unquestionable facts. These were now to be provided by Xenophanes (570–475 B.C.), a one-time resident of the Ionian city of Colophon.

EMERGENCE OF SCIENTIFIC SPIRIT

Xenophanes' main interest, however, was not to provide an objective foundation for the support of Greek science. What he was really concerned about was what he considered to be the mythological nonsense in Greek religion. Xenophanes was in fact, more a theologian than a scientist. It was only to place his own thoughts above mere speculation that he deemed it important to introduce irrefutable evidence in support of his ideas.

Xenophanes first took issue with the contemptuous portrayal of the gods by Homer and Hesiod: "Homer and Hesiod have attributed to the gods everything that is a shame and reproach among men, stealing and committing adultery and deceiving each other."[1] In Xenophanes' opinion, the source of this contempt for the gods was nothing short of the disregard that these poets felt for their fellow man; the behavior ascribed to the gods was actually a reflection of their own attitudes and behavior.

Xenophanes' next volley was aimed at the anthropomorphizing tendencies in man's religious thinking: "The Ethiopians say that their gods are snub-nosed and black, the Thracians that theirs have light blue eyes and red hair . . . if cattle and horses or lions had hands, or were able to draw with their hands and do the works that men can do, horses would draw the forms of the gods like horses, and cattle like cattle, and they would make their bodies such as they each had themselves."[3] In other words, each race sees its gods in its own image.

Xenophanes' contention was that the gods or god was "in no way similar to mortals either in body or in thought."[4] It was time for men to abandon their distorted pretensions, for the gods as man had formerly pictured them, did not exist. While he conceded that there were many gods, Xenophanes maintained that of these there was one supreme god who "'always remains in the same place, moving not at all . . . but without toil he shakes all things by the thought of his mind."⁵ "All of him sees, all thinks, and all hears."⁶

While it was certainly not original to argue that there was one supreme god in the pantheon of gods, the idea that an intelligence could translate its will into deeds without the use of a body was something entirely novel to the Greek way of thinking. Beginning with this new cognition of the divine, Xenophanes then set out to discover for himself the basic material comprising the origin of all things.

Xenophanes eventually came to the conventional decision that "all things that come-to-be and grow are earth and water,"⁷ but unlike those who had come to the same conclusion, Xenophanes offered compelling evidence to substantiate his point of view.

This evidence took the form of fossils, discovered in some very unexpected locations. For example, sea shells had been found inland, even on mountains, and rocks bearing the impression of fish had been discovered in the rock quarries of Syracuse. In Paros, the impression of a bayleaf was found in a rock, while in Malta, the outline of marine objects had been detected in the mud.⁸ These facts had been known long before Xenophanes came upon them, but it was Xenophanes who first realized their significance and meaning. To Xenophanes the discovery of these fossils in such remote surroundings could mean only one thing: the earth had once been covered by the sea and plants and animals had come to life from the mud and rocks when these had emerged out of the water. The theory of Abiogenesis now had a sound empirical foundation.

IV

Parmenides of Elea (fl. 450 B.C.) is the next important philosopher to contribute to the development of evolutionary thought. His contribution was, however, indirect, for it was actually as a reaction to his views that many subsequent philosophers made the advances they did.

Unlike his predecessors, Parmenides was not born in Ionia, a factor that may explain his radical departure from the tenets of Ionian philosophy. His birthplace was Elea, a colony situated on the coast of Lucania, to the south of Poseidonia. It was here that he founded what was to become the Eleatic "school," which was to dramatically change the course of Greek philosophy.

Hitherto, there had been no opposition to the Milesian concept of a primary substance. Thales, Anaximander, Anaximenes, and Heraclitus had each argued that the world was made up of varying amounts of a single basic substance, even though they could not agree as to its composition. Xenophanes had even adduced paleontological evidence to support their argument for a primary substance as the basic material of life. Parmenides, however, was not convinced.

In his opinion, to acknowledge a primary substance and a postulate that stated that objects differed because there was more of the primary substance in one place than in another, as had Anaximenes, would mean that one would also have to accept the possibility that in some places a condition of "'Not Being" also existed. This in turn would mean that a void, i.e., "Not Being," could be transformed into matter, i.e., "Being," since Anaximenes had argued that matter could pass from one extreme of the continuum (of air) to another.

Parmenides refused to accept this approach, "for never shall this be proved, that things that are not, are."[1] What Parmenides was in effect saying was that the concept of "Being" precludes that of "Not Being." Moreover, he felt that such a thing as "Not Being" could not exist, and therefore pure space, which is "Not Being," also could not exist. Furthermore, since the concept of change requires space, it too was fundamentally unsound.

In pursuing this line of thought Parmenides was forced to the conclusion that changes in matter do not occur, despite sensory evidence to the contrary. Man's senses had deceived him. The only source of truth was reason and to Parmenides, reason dictated that it was impossible for matter to undergo change.

The hypothesis that all matter was essentially composed of more or less a single primary substance was therefore also untenable, for there was no means by which material could be added to or taken from the original substance. Since one kind of matter could not change into another, matter must not be made up of differing amounts of a single substance, as the Ionian philosophers had contended. The only alternative solution to the variations that existed in the world was that matter was composed of a number of basic materials. Here was born the doctrine of pluralism.

This doctrine was to have great impact on Parmenides' contemporaries. They were impressed by his refutation of the monistic hypotheses and his cogent denial of the possibility of change. Had Parmenides' arguments regarding the superiority of reason over the senses also been accepted, the scientific movement would have stagnated at this point. Fortunately, however, most of his contemporaries did not endorse his denial of sensory evidence as a valid means of gaining information.

Parmenides' main contribution to the growth of science was thus the direction he gave to subsequent cosmological theories. Matter was no longer thought to be more or less of a certain quantity of a basic element. The new understanding was that of a conglomeration of several basic elements, differences arising as a result of varying proportions of these materials.

Changes and differences were henceforth thought to be due to a rearrangement of these substances.

V

Empedocles of Acragas (495–431 B.C.) was one of the first philosophers to be influenced by Parmenides' pluralistic doctrine. But at the same time he felt that each of the first principles proposed by the Ionians as essential elements in the composition of matter also warranted attention. The only solution was a compromise—both positions had indisputable merit and he sought to combine them. The result was the postulation of four initial substances: "shining Zeus, life-bringing Hera, Aidoneus and Nestis who with her tears fills the springs of mortal men with water";[1] in other words, fire, air, earth, and water.

To activate these four elements (for motion was no longer thought to be an innate characteristic of matter), Empedocles postulated the existence of two basic forces: Love and Strife. The former was a uniting force, the latter, divisive. With these two forces, or sources of energy, as it were, and the raw material of fire, air, earth, and water, Empedocles then began to describe the origins of life in a manner rivaling, if not outdoing, the imagination of the lurid artist of the Middle Ages, Hieronymus Bosch.

Four stages of evolution were proposed. First was the stage marked by the appearance of creatures with disunited limbs: "The first generations of animals and plants were not complete but consisted of separate limbs not joined together."[2] There were "faces without necks, arms wandered without shoulders, unattached, and eyes strayed alone, in need of foreheads."[3] . . . Limbs wandered alone."[4]

The second stage, "arising from the joining of these limbs, [was] like creatures in dreams."[5] But as one divine element mingled further with another, these things fell together as each chanced to meet the other, and many other things besides these were constantly resulting."[6] Many creatures were born with faces and breasts on both sides, man-faced ox-progeny, while others again sprang forth as ox-headed offspring of men, creatures compounded partly of male, partly of the nature of female, and fitted with sterile parts."[7]

These two stages occurred during the period in whch Love was in the ascendancy and Strife was kept from exerting its influence. The rise of the limbs and their random unions were both caused by the force of Love, which brings together all things with no regard for uniformity. The next period was governed by Strife, which managed to gain the upper hand. What Love hath joined, Strife now attempted to put asunder. But this was not an indiscriminate disunion, for "wherever . . . everything turned out as it would have if it were happening for a purpose, there the creatures survived, being accidentally compounded in a suitable way; but where this did not happen, the creatures perished and are perishing still."[8] Possibly Empedocles was noting at this point that the capacity to reproduce was an important prerequisite for those creatures which were to survive the second period—a capacity not characteristic of those creatures which ultimately disappeared.

The infertile monsters perished and in their place arose "whole-natured forms . . . showing forth as yet neither the lovely form of the limbs, nor the voice nor the organ proper to men."[9] In other words, these creatures were sexless and could not be distinguished as to species. They were in essence undifferentiated source material.

In the fourth stage, which was also governed by Strife, differentiation occurred through a separation of parts from the whole, and many species of both sexes were brought forth. Thereafter, all unions between two creatures involved the attraction of like substances. "Thus, sweet seized on sweet, bitter rushed towards bitter, sour moved towards sour, and hot settled upon hot."[10]

The fourth stage had not ended, however, for, as Empedocles saw it, it was still going on and changes were still occurring. For example, even though the first animals were all born with spines that were straight, some animals now had rounded spines because certain of their ancestors adopted the habit of turning their necks backward for some reason.[11] This caused their spines to bend and resulted in an alteration of their physical characteristics. Presumably, this acquired characteristic was passed on to their offspring with the result that some animals now had rounded spines. In like manner, Empedocles noted that "in the realm of animals they become lions that have their lair in the mountains, and their bed on the ground; and in the realm of fair-tressed trees, they become laurels."[12] Evidently, he anticipated Lamarck in considering that habits and environments could potentially influence both morphology and phylogeny. Man himself is not the same creature that he was when he first appeared, for "in so far as their natures have changed (during the day), so does it befall men to think changed thoughts (in their dreams)."[13] In other words, during sleep, man's thoughts reveal his original characteristics through the very content of his dreams. "For from these (elements) are all things fitted and fixed together, and by means of these do men think."[14]

Having surveyed Empedocles' theories on the origins of life, one cannot fail to feel a certain astonishment at how closely he approximated some of Darwin's own principles of evolution. For instance, in his explanation of how the curved spine and the lion came to be as they are, Empedocles appears to be introducing a rudimentary version of Darwin's own theory of

Natural Selection. And, in fact, these evolutionary principles were correctly perceived and rephrased even in his own age.

For instance, Simplicius[15] explained the gist of Empedocles' theory to his readers in the following words: "As many of these parts that were fitted together in such a way as to ensure their preservation, became animals and survived, because they fulfilled mutual needs—the teeth tearing and softening food, the stomach digesting it, and the liver converting it into blood. The human head, when it meets a human body, ensured the preservation of the whole, but being inappropriate to the ox-body it leads to its disappearance."

For his part, Aristotle[16] was taken by the manner in which Empedocles had introduced the idea of purpose in Nature without having recourse to teleological principles. One might even see Aristotle's criticism of Empedocles' theory in light of the doctrine of the survival of the fittest: "Where everything turned out in a way which simulated purpose, these creatures survived because by chance they were constituted in a certain way; whereas all that were not so constituted perished, and continued to perish . . . just like the oxen with human faces."

Empedocles, however, did not teach the doctrine of evolution. As H. F. Osborn correctly noted, Empedocles can be regarded as an evolutionist "only in so far as he taught the gradual substitution of the less by the more perfect forms of life."[17] Empedocles did not, in fact, advance the notion that less perfect forms were *succeeded* by more perfect forms; his contention was that they were *replaced* by them. In essence, his theory was still an account of the origins of life in terms of spontaneous generation: life arose spontaneously as a result of the action of energy (Love and Strife) upon four basic elements. First came the imperfect forms of life, which later disappeared not because of their grotesque appearances, but because they were unable to reproduce. When these creatures

EMERGENCE OF SCIENTIFIC SPIRIT 41

perished, they were replaced by forms that had greater survival potential. Of these, those which could adapt themselves to their environments survived and flourished; those that could not disappeared as had their predecessors. Yet, because of his insight into the processes of Natural Selection, and because he had, in fact, a premonition of an evolutionary unfolding, one might be inclined to agree with Osborn in calling Empedocles, "'the father of the idea of Evolution."[18]

VI

After the conquest of Ionia and the destruction of Miletus in 494 B.C. by the Persians, it was Athens that emerged as the center of Greek commerce and learning and, therefore, it was to Athens that many of the leading philosophers of the period gravitated. But despite its democratic government, Athens was also a conservative city, jealous of its traditions and proud of the divine protection afforded it by the gods. Many Athenians would simply not tolerate any disguised denial of the gods, as was occurring throughout the rest of Greece. And so, when the Ionian philosopher Anaxagoras (500-428 B.C.) was overheard teaching that the sun was not really a god but only a red-hot stone,[1] many of the citizenry became incensed and he was imprisoned. Had it not been for his friend Pericles, who arranged for his escape and passage back to his native Ionia, Anaxagoras would surely have been put to death by the Athenians, just as Socrates was a few years later.

However, Anaxagoras is most renowned not for his nearmartyrdom, but for his postulation of *Nous* (Mind); a spirit or force separate from matter and responsible for bringing the world into its present state.

The *Nous* was Anaxagoras's solution to the regularities found in Nature: Nature behaved according to strict laws because it had been designed to act that way. To support this notion, Anaxagoras resorted to the analogy between the *Nous* of man and the *Nous* of the universe. Just as man had a mind that could bring order out of a chaotic arrangement of facts, so too the universe must have some guiding force, for how else could the uniformities and laws of Nature be explained?

This insistence that the *Nous* was responsible for what appeared to be orderly and purposeful in life, is Anaxagoras's most important contribution to evolutionary thinking. Matter and life did not arise as if by chance, but by design. From this premise, Anaxagoras then went on to spell out the details of creation.

The first thing he had to deal with, however, was Parmenides' arguments that there could be no such thing as origination or destruction. The answer was that the basic building blocks of matter were preexistent: "bones are formed out of extremely small, minute pieces of bone—flesh is formed out of extremely small, minute pieces of flesh; blood is formed when many droplets of blood come together."[2] No other solution was conceivable, for "how could hair come from what is not hair, or flesh from what is not flesh?"[3]

But given that the "seeds" of everything were always in existence, it was still necessary to explain differences and similarities between the various forms of matter in the world. The Milesian monistic philosophy, which accounted for qualitative differences in terms of the quantity or prevalence of a single basic substance, was no longer tenable, owing to the force of Parmenides' argument for the pluralism of basic elements. And since nothing came into being or disappeared, the only possible explanation was that some of the preexistent substances had been made to mingle and combine while some had not. Thus, "the dense and moist and cold and dark (elements) collected here, where now is Earth, and the rare and hot and dry went

EMERGENCE OF SCIENTIFIC SPIRIT 43

outwards to the furthest part of the Aether."[4] The seeds of plants and animals floating in the air also formed through these early combinations, after which they fell to the earth with the rain.[5] Having once generated, they became capable of duplicating themselves through the usual means of production.[6] All this was initially set in motion by the *Nous*.

Thus, for Anaxagoras, there was no spontaneous bursting forth from the earth or water. The seeds of all life-forms were preexistent and required only some external force to cause them to mingle and generate. Empedocles had seen such a force in the conflict between Love and Strife; for Anaxagoras it was *Nous*, and, because of *Nous*, there was order in the universe.

VII

Anaxagoras's imaginative solutions to Parmenides' arguments against a creative process had been that a) the seeds of all matter had existed from the beginning of time, and b) Mind was the force or energy that initially caused them to germinate. But these were not the only solutions to be proposed. The Atomists, headed by Leucippus (fl. 430 B.C.) and Democritus (fl. 420 B.C.) of Abdera in Thrace, took a somewhat different approach to these problems.

For them, the fact of motion required no metaphysical explanation. Like the Milesians before them, they contended that motion was an innate characteristic of matter and that matter itself had always been in existence. The theory that these philosophers then introduced was a combination of the best of Milesian and Parmenidean philosophy.

First, they advanced the principle of the conservation of matter: Nothing can be created out of nothing, nor can it be destroyed and returned to nothing."[1] However, matter could take different forms and could be separated into various basic substances. Yet there was a point beyond which this process of division could not be extended, and this finite limit was the atom.

The atom was an indivisible, corporeal body and a unit unto itself. It could undergo no changes; it could not come into being or cease to be. It could only unite with or separate from other atoms. Atoms were "infinite both in number and in the varieties of their shapes, of which everything else is composed— the compounds differing from one another according to the shapes, positions, and the groupings of their (atomic) constituents."[2]

The second major principle that the Atomists proposed was that, contrary to Parmenides, empty space did exist and, as a result, atoms were able to move. The existence of space was a logical necessity, for "unless there is a void with a separate being of its own, 'what is' cannot be moved—nor again can it be 'many,' since there is nothing to keep things apart."[3]

With the assumption of these two ideas, the Atomists could now account for the differing kinds of matter found throughout the universe: By virtue of their own innate characteristc of motion, the atoms move in the infinite space, colliding into each other and bouncing off one another until some become "intertwined one with another according to the congruity of their shapes, sizes, positions, and arrangements," so that some "stay together and so effect the coming into being of compound bodies."[4]

With more and more of these compound bodies being formed, countless worlds might come into being, for in truth "there is no end to the universe, since it was not created by any outside power."[5] Thus, "in some worlds there is no sun or moon, in others they are larger than in our world and in others more numerous" and "there are some worlds, devoid of living creatures or plants or any moisture."[6]

Although neither Democritus nor Leucippus seems to have directly concerned himself with the appearance of man or the animals, it seems from the last fragment of their writings that they envisioned the appearance of plants and animals in the same way as they envisioned the appearance of inanimate matter, i.e., chance and chance alone was responsible for the different species of animals.

In essence, this conclusion means that the Atomists followed their predecessors in believing in a kind of abiogenesis—by chance certain atoms came together and, as a result, living organisms emerged. Although we have no evidence, we must assume that a process akin to Empedocles' imperfectly and perfectly formed creatures came into being and that only the latter were capable of survival. The others perished, their atoms separated, and they reentered space to possibly take up other forms of matter.

Considering the infinite possibilities that might arise through the chance collisions of different atoms, the atomic theory could account better than any other prior theory for the manifold species of plants and animals that the Greeks observed in Nature. Furthermore, given that the atom had always been in space and that motion was an intrinsic property of its being, there was no need to posit an outside force to explain design or purpose in Nature. Chance, and chance alone, was responsible for what appeared to be the product of intelligent design.

Owing to Aristotle's influence, however, the Atomic theory never gained the recognition it deserved, except in the philosophical systems of Epicurus and Lucretius, which I shall discuss somewhat later. Aristotle's main criticism of the school stemmed from his disagreement with the Atomists over the source of motion. Thus, in his *Metaphysics* Aristotle wrote that Leucippus and Democritus "say that there is always movement. But why and what this movement is they do not say, nor, if the world moves in this way or that, do they tell us the cause of its doing so. *Now nothing is moved at random, but there must always be something present to move it; e.g., as a matter of fact a thing moves in one way by nature, and in another by force or through the influence of reason or something else"* (italics mine).[7] "When therefore Leucippus and Democritus speak of the primary bodies as always moving in the infinite void, they ought to say with what motion they move and what is then natural motion."[8] Anaxagoras had stated that the source of motion was an outside force called Mind. For this, Aristotle designated him a "sober man" as compared to his predecessors and contemporaries.[9]

But in so doing, Anaxagoras had introduced the Mind-body problem into science. The Atomists, however, refused to be trapped in this metaphysical snare, and ultimately, it was their approach that received empirical support in the form of Brownian movement.[10] Here, with the aid of the microscope, can be seen the random purposeless movement of atoms placed in a liquid or gas, exactly as the Atomists had envisioned.

Before leaving the Atomist school, however, we must note the important concession still made by Democritus, and later by Plato, to Parmenides, namely, that pure reason and not observation was the ultimate means to knowledge. Thus Democritus wrote, "There are two sorts of knowledge, one genuine and one obscure. To the obscure belong all the following: sight, hearing, smell, taste, touch. The genuine is separated from this. . . . When the obscure can do no more—neither see more minutely, nor hear, nor smell, nor taste, nor perceive by thought—and a finer investigation is needed, then the genuine comes in as having a tool for distinguishing more finely."[11]

It was this negative attitude toward the evidence of their senses that eventually put a limit to the achievements of Greek science, for in denying the evidence of their senses, the Greeks never thought to test their theories experimentally. As a result, their science stagnated in philosophical argument.

VIII

A number of tentative hypotheses had now been proposed in response to the questions of the *arche* and the origins of life. The last decades of the pre-Socratic period were to witness the difficult task of synthesizing these various solutions into an acceptable theory. The two leading figures in this trend toward eclecticism were Diogenes of Apollonia and Archelaus of Athens.

Diogenes' (fl. 440 B.C.) starting point was the hylozoistic principle of the Ionian and Atomist philosophers: matter contained its own energy. Second, he rejected the pluralistic doctrine of Parmenides and his followers and instead reintroduced Anaximenes' general proposition that Air was the *arche* and that everything else arose from quantitative changes in the composition of this substance.[1]

The reason for Diogenes' endorsement of Ionian monism was his conviction that qualitatively different materials would not be able to coalesce. Therefore, they would be unable to unite to form new substances: If "earth and water and air and fire and all other things apparent in this world-order were different from the other" it would not be possible for "a growing plant to grow out of the earth or for a living creature or anything else to come into being, unless they were so composed as to be the same thing. But all these things being differentiated from the same thing (i.e., air), they are differentiated and take different forms at different times, and return again to the same thing."[2]

Diogenes, however, was not a complete reactionary. Although he subscribed to the monistic doctrine, he also endorsed Anaxogoras's concept of Nous, which he equated with his first principle, air: "And it seems to me that that which has intelligence is what men call air, and that all men are stirred by this and that it has power over all things."[3] This had to be true because "it would not be possible without intelligence for it (air) to be divided up so that it has measures of all things."[4] As further proof, Diogenes pointed out that "men and the other living creatures live by means of air, through breathing it. And this is for them both soul and intelligence . . . if this is removed, then they die and intelligence fails."[5]

Turning to the problem of the different forms of living things, Diogenes went beyond Anaximenes' basic premise that animals were merely the product of different degrees of rarefaction and condensation. In his estimation, the distinction between animals also had to include differences in internal temperature as well: "in none of living creatures is this warmth alike; nor, for that matter, in any two men; but it does not differ much, only in so far as is compatible with their being alike. At the same time, it is not possible for any of the things which are differentiated to be exactly like one another till they all once more become the same."[6]

Temperature changes were also significant in the theories of the other important eclectic thinker of this age—Archelaus of Athens. According to Diogenes Laertius, Archelaus had been one of Anaxagoras's students and it is also reported that Archelaus was once mentor to Socrates.[7] According to various traditions he is also said to have taken over leadership of the school at Lampsakos after Anaxagoras's death. Judging by his philosophy, however, he seems to have followed the Ionians to a greater extent than he did his teacher. For instance, he is reported to have taught that "water is melted by heat and produces earth in so far as by the action of fire it sinks and coheres, while on the other hand it generates air in so far as it overflows on all sides."[8]

It is in his views regarding the origins of life, however, that Archelaus was to make his main contribution. The analogy of Mother Earth and human motherhood had always attracted the imagination of the Greeks.[9] Archelaus appears also to have been captivated by the similarities, so much so, in fact, that he proposed that when life was first emerging from the earth, a milk-like fluid was also produced so that these first creatures would have a source of nourishment to carry them through their early period of development. Diogenes' account of Archelaus's hypothesis is as follows: "Living things are generated from the earth when it is heated, and it throws off slime consisting of milk to serve as a sort of nourishment; and in this same way the earth produced man."[10]

The only other record of Archelaus's theory is that preserved by Hippolytus: "As to animals, he (Archelaus) says that when the earth was first being warmed, in the lower part where the warm and cold were mingled together, many living creatures appeared, and especially men, all having the same manner of life, and deriving their sustenance from the slime; they did not live long, and later on generation from one another began."[11]

This analogy between Mother Earth and human motherhood was now enhanced by Archelaus's introduction of a primordial milky fluid, and was subsequently plagiarized by most cosmologists of the period.[12] Here is Dio Chrysostom's borrowed account of man's early days: "As the first nourishment the first men, being the very children of the soil, had the earthy food—the moist loam at that time being soft and rich—which they licked up from the earth, their mother as it were, even as plants do now draw their moisture therefrom."[13] Many others were to write in a similar vein.[14]

3
Pre-Socratic Scientific Accomplishment

Now that we have considered the cosmological speculations of Antiquity up to the time of Diogenes and Archelaus (the last of the important pre-Socratic philosophers), it seems appropriate to pause and evaluate the changes in thinking that occurred during this seminal period in the history of ideas.

The first of these changes involved man's assumptions about his gods. Prior to Thales, the gods were deemed responsible for whatever occurred in Nature, and since the minds of these gods were held to be inscrutable, no one felt competent enough to inquire into their natural domain. Consequently, the discovery of probable causation in Nature could never occur until the possibility of divine intervention was dismissed. Unless the gods ceased to occupy an important position in the explanation of Natural phenomena, there would always be the outside chance that some capricious god was responsible for the event. Only with the assumption that the gods did not play any role in the regulation of the natural order of the universe, could man feel that his efforts to understand Nature were not in vain. It was this important assumption which set the Greeks off from all the other peoples of Antiquity.

The second unique development in thinking was the attempt to explain the diversity of substances in the world on the basis of a transformation of a few basic materials or elements. The various materials that were considered as primary were water (Thales), fire (Heraclitus), and air (Anaximenes). More profound was the hypothesis that all substances were present from the beginning of time, but were combined in such a manner as to make them indistinguishable. Matter only came into being when a separation of these substances ultimately came to pass (Anaximander). Somewhat related to this theory was the notion that all things were present from the beginning in miniature ("seeds"), and that matter arose when these seeds eventually germinated (Anaxagoras). Then there was the theory of a limited number of primordial elements, such as earth, fire, water, and air, which were always present and which combined to produce all the other known substances (Empedocles). And finally, there was the Atomic theory, which taught that there were an infinite number of elements that were always present and that combined by chance to produce all the other known substances (Leucippus and Democritus).

Although the pre-Socratic philosophers were patently superficial in the generalizations they drew from their observations, the very fact that they did venture generalizations is to be reckoned as another important step in human thought. It is this capacity for abstraction and inference, and not the scientific validity of their theories, which is significant in the history of ideas. Without this trend toward abstract thinking and generalization—this tendency to subjectively organize facts into higher-order concepts—man could never hope to understand his factual environment; he would forever be tied to individual events and no meaningful interpretation of his observations would be possible. The facts themselves are secondary. It is the analysis of these facts that is the beginning of understanding. "The really important thing in science is the idea, once given, it must be left to work itself out in practice."[1]

The following examples indicate how close the pre-Socratic philosophers came to anticipating some of the basic principles of science in the twentieth century. Again, it is their ideas and not their facts that are of importance.

For example, some of the pre-Socratic philosophers argued that life arose from the primeval waters and that water was the basic component of life. Today there is general agreement that life arose from an archaic sea, and water has long been recognized as the most fundamental component of the body.

Although modern biology does not acknowledge the pre-Socratic doctrine of Abiogenesis, it does endorse the notion that the organic matter of living tissue must originally have arisen out of nonliving matter. With respect to the modern theory of evolution, we have also noted that many of the principles of our own theory were anticipated by Empedocles. Among these were the ideas that less perfect forms appeared before more perfect, "whole-natured" forms, and the idea that there was a process akin to natural selection that determined the eventual behavior and physical characteristics of the various species of living organisms.

The pre-Socratics also taught the principle that the primary substance or substances of which all matter was composed could be neither created nor destroyed. Substituting the idea of energy for matter, one may then recognize the same principle in modern physics concerning energy, that it too can be neither created nor destroyed. In other words, energy (=Nous, Love and Strife) has always been present.

The high point of pre-Socratic philosophy is, of course, the Atomic theory. It is today the main working hypothesis of the science of chemistry, and it is also recognized as a valid statement of how life must have originated in the archaic sea.

Given that these ideas came so close to approximating the theories of contemporary science, it does not seem possible that they are separated by a period of two thousand years and more. Why did it take so long for modern man to catch up to Antiquity? The answer lies in the turn taken by philosophy in the generations that succeeded the pre-Socratics. In place of looking outward, man began to examine himself; in place of observation came self-awareness, self-criticism, and self-evaluation.

4
War and the Demise of Greek Science

Before the fifth century B.C., Ionia had been the recognized center of philosophy and science in Greece. But the Persians eventually began to exert political pressure upon the Greek city-states in Asia Minor, and, as hostilities between the two peoples became more and more imminent, many of those who could afford to, left and migrated to mainland Greece. With this sudden influx of writers, artisans, philosophers, and the like, the city-state of Athens soon emerged as the new "school of Hellas."

The Persians, however, were bent on expansion, and they did not stop at the conquest of Ionia. At first they marched over mainland Greece with little opposition, but first at Marathon (490 B.C.) and then at Salamis (480 B.C.) they suffered defeat at the hands of a Greek contingent half their size. The Persians were ultimately forced to withdraw, leaving Greek independence still intact.

At the forefront of those who opposed and eventually routed the Persians were the Athenians. And now that they were victorious, they became heady with self-confidence. The city was at the height of its glory and its citizens felt it. There was an atmosphere of excitement in Athens that attracted sightseers from all over Greece, and many of those who had planned

DEMISE OF GREEK SCIENCE 55

to come there for only a short visit, ended by settling permanently.

Athens began to grow more and more prosperous. A cultural and economic empire was being built and every other city-state in Greece was gradually falling under Athenian domination. It appeared that Athens was about to emerge as the foremost political power in Greece as well. Sparta, however, refused to yield to Athenian domination, and the tension between the two city-states soon erupted into the Peloponnesian war in 431 B.C. The fighting continued off and on for the next twenty-seven years. By 404 B.C., however, Athens was too exhausted to continue. Not only had the city suffered the miseries of war; it had also experienced the terrifying agonies of a plague that had ravaged Athens for five whole years during the height of the fighting. Its manhood spent, Athens was forced to accept defeat.

The humiliation and the inevitable after-effects of war effected a marked change in those Athenians who survived. The rule of the *demos* became the rule of the mob, and the internal affairs of the city fell into a general state of chaos. The farmlands had been devastated, the economy was disrupted, and everywhere there were ex-soldiers who could not make the readjustment to civilian life.

An aristocratic party known as the Thirty Tyrants then took control, and a reign of repression and conservatism was imposed on the city. Corruption was rampant and criticism was stifled. Anyone even suspected of opposition was subject to arrest. It was during this period that the famed philosopher Socrates was sentenced to death on a charge of impiety—denying the existence of the national gods. The actual reason for his sentence, however, was probably his criticism of the ruling powers.

This was the atmosphere in which Plato (430–347 B.C.) grew to manhood. Born the son of a wealthy Athenian aristocrat, the excesses and intolerance of the aristocratic party dismayed him, and instead of a political career, he chose the quiet unassuming path of the philosopher.

Growing up in Athens, he was attracted to Socrates and studied with him until the latter's death. Thereupon he was forced to flee Athens for a time, for to be a friend of a convicted man was somewhat dangerous. Plato did not end his studies, however. Instead, he journeyed to Megara, which lay between Athens and the city of Corinth, and there he delved into Eleatic philosophy with his friend Euclid.

It may be remembered that Parmenides was in large measure the founder of the Eleatic school, and it was Parmenides who had taught that sense information was an unreliable source of knowledge. Hence, it is not surprising that when Plato returned to Athens around 390 B.C., he began to teach that the only true knowledge was that which came from reason.

The belief that perception gave a veridical account of reality was unacceptable, he argued, because what appeared to be true for one individual often did not appear so for another. Observation and experimentation could only give one an opinion. Plato felt that knowledge had to be the same for all. It could not be subject to distortion; it had to be stable, universal, and permanent. Since the senses made the world appear unstable and ever-changing, sensory information had to be distrusted. The only valid means of attaining truth was through reason.

Unfortunately, Plato was to exercise a lasting influence not only on his own pupils but on nearly all those students of the next generation and beyond, who were compelled to read his works as an integral part of their own education.

Acceptance of Plato's philosophy had a predictable effect. The inclination toward observation was undermined, and in its place came a devotion to the solution of abstract problems. Experimentation ceased. Would-be inventors were now loath to devote their time to activities that would only be denigrated by their teachers. The scientific tradition that had been sown and nurtured among the Greeks was beginning to wither. Death would soon ensue.

A pertinent example of Plato's rational approach to matters scientific was the theory of evolution that he proposed in the "Timaeus." Man, he stated, had been fashioned out of the four basic elements, earth, air, fire, and water, by a deliberate and creative deity and his aides, the lesser deities. Into the body that had been thus formed, the supreme deity placed a rational soul in its head and heart. Into the remainder of the body, the lesser gods placed an irrational soul. The farther from the head, the less rational it became. The body itself became the battleground for the opposite tendencies of these two souls and the behavior of the individual was a direct reflection of the tendency that became dominant.

The soul, however, was immortal. After death it did not perish, but instead was either elevated or degraded to a different niche of existence, depending on the individual's behavior while he was alive. Those who had lived the life of the philosopher had their souls exalted to the world of the celestial soul—the world of thought. Those who had lived a less worthy life had their souls descend to the lower world of the animals. The type of animal into which the soul descended reflected the individual's way of life while he was still alive, and a special animal body was made to accommodate that particular soul.

Thus, birds were said to come from dim-witted but harmless men who had watched the heavens in their belief that knowledge might be obtained solely through the evidence of their eyes. Those who became land animals were those who had had no use for philosophy and who had given vent to their instincts instead of their reason. Finally, there were the fish—the incarnation of the souls of those who had been foolish and ignorant while alive. They were held to be unworthy even of breathing air and so they were placed in the water away from the other creatures of the earth.

Ludicrous though they were, the ideas Plato put forth in the "Timaeus" were to exercise a profound impact upon the progress of science in the West. Especially is this true of his notion of a dualism between body and soul, with the former serving only as an encasement for the latter. The body-soul dichotomy gained ready acceptance among many of his students, and the picture of the animal world as the visible reflection of the inner world of man was to be developed and made the working hypothesis of the theory of evolution for the next millennium by Aristotle, the most eminent of all the Greek philosophers.

5
Aristotle's Views on the Origins of Species

Few individuals can be compared to Aristotle in terms of his influence on Western thought. He was the first great encyclopaedist and his observations, real or imagined, stamped the course of Western science for hundreds of years after his death.

Born in 384 B.C. at Stagiras, a Macedonian city about 200 miles north of Athens, Aristotle was introduced to court life at an early age, since his father was personal physician to Amyntas, the king of Macedonia. Following his father's death in 367 B.C., Aristotle made his way to Athens to further his education. There he fell under Plato's spell and he studied at the Academy for the next twenty years of his life.

In 347 B.C. Plato died and Aristotle left Athens to take a position with Philip of Macedon as personal tutor to the king's son, Alexander. He did not remain long at this post, however, for Alexander soon afterwards embarked upon his conquest of the known world.

In 342 B.C. Aristotle was back in Athens and this time he founded his own academy, the Lyceum. Here he remained until 323 B.C., the year in which Alexander died. The great conqueror had made enemies of many of the inhabitants of Greece and because of Aristotle's association and friendship with Alexander, the resentment that the latter had created was turned against those who had been partial to him. Accused of impiety, the same charge that the earlier Athenians had used to eliminate Anaxagoras and Socrates, Aristotle fled for his life to Chalcis in Euboea. Shortly thereafter, he died a quiet death.

Aristotle was one of the most prolific writers of the ancient world. His estimated writings number almost a thousand, but of these only a fraction has been preserved. In these extant writings, the philosopher reviewed all the known facts of his day and then he proceeded to formulate his own interpretation of what these meant.

Although Aristotle never went into the question of evolution with any clarity, he did clearly side with Anaxagoras against Empedocles and the Atomists in arguing that there was purpose in Nature. This purposefulness could be seen in the gradual progression of life from plants through animals to the ultimate biological specimen—man. The attainment of this end was the main plan in Nature. Plants and animals, like women and monsters, were simply failures that had come about because Nature had to contend with "'slipshod" working material.[1]

Aristotle's thoughts on the evolution of these imperfect forms is caught up in his treatment of the soul as the designing principle in Nature. Although these notions regarding the influence of the soul were radically different from Plato's, the influence of teacher on pupil is readily apparent.

The form of an organism and the functions that that organism would be capable of performing were determined by the type of soul contained within it. Observation of the behavior of the various forms of life indicated that there were three basic types of soul. The most primitive and therefore that shared by all life forms was the nutritive-generative soul. This was the soul possessed by plant life and, because of its qualitative simplicity, the only movement available to plants was upward and the only method by which they might reproduce was by seed.

But from whence did these primitive forms arise? They were preexistent, according to Aristotle. Life containing matter did not develop from lifeless matter but was present within it from the beginning of time. It had required only moisture and a suitably high temperature to make it appear.[2]

The second and more highly developed type of soul was that which imparted sensation. Possesion of this soul in addition to the nutritive-generative soul of the plant world was the distinctive feature of the animal world. A body with such a soul was one that could move in its environment so as to change the pattern of sensations to which it might be exposed. A body with this type of soul meant that it could interact with other bodies and hence sexual reproduction was possible among animals.

But though animals shared the nutritive-generative soul with plants and though they resembled plant life during their own embryonic stage, Aristotle did not regard animals merely as highly deveoped plants. True, Nature proceeds little by little from things lifeless to animal life, in such a way that it is impossible to determine the exact line of demarcation,[3] but this is not to be taken as evidence that one evolved from the other. Just as plants had come into being, so too animals had spontaneously erupted from "putrefying earth or vegetable matter, as was the case with a number of insects while other are spontaneously generated in the inside of animals out of the secretions of their internal organs."[4] But not only did insects appear in this way. Aristotle also adduced as evidence a rather common observation among naturalists, namely, "that certain fishes come spontaneously into existence, not being derived from eggs or from copulation,"[5] a reference, no doubt, to that class of fish known today as mouth-breeders. The remaining animals that did not emerge spontaneously and fully grown from lifeless matter, arose from eggs or from a scolex,[6] and then grew into their eventual natural forms. But from their very moment of existence, each possessed a characteristic soul that would determine its development and behavior.[7]

The ultimate attribute that a soul could impart was reason. This characteristic was possessed by man alone, in addition to the sensitive and nutritive-generative souls of the other forms of life. Man was thus the most highly developed being, since his body, and his alone, contained the type of soul that conferred wisdom.

Since Aristotle saw reason as the function that rendered man the supreme creature, one may infer that he regarded the attainment of reason as the goal toward which life was directed. While man represented the apogee in the progression from nonreasoning to reasoning matter, there was still one step beyond even this level, and that was pure reason—Aristotle's concept of God.

For Aristotle, pure thought was God. It could be no less. To think of man—his accomplishments, his motives, his goals—this was the realm of man. But to think about thinking—that was God's domain. And this "thinking about thinking" was Aristotle's idea of God.[8] As such, God had no form, nor could God enter into any activity outside the realm of thought. God was a self-contained entity, totally aloof from life on earth. Hence, God could not have had a hand in the appearance or the development of life.

Plants and animals had sprung from the earth. Where did man come from? Did he develop from the other forms of life? Apes and monkeys, he noted, shared some of man's nature: "as regards man and animals, certain psychical qualities, e.g., cunning, courage, timidity, etc. are identical with one another,

Abiogenesis had met its first formidable critic, but it was not until well into the present era that the hypothesis was finally laid to rest.

It has been suggested by some historians of early biology that Theophrastus predated Lamarck in urging that acquired characteristics could be inherited and passed on to offspring in plants.[3] Theophrastus's own statement was that "the soil seems to produce plants which resemble their parents; on the other hand a few kinds in some places seem to undergo a change, so that wild seed gives a cultivated form or a poor form one actually better," however, "this is not a change, but a natural development towards a better or an inferior form."[4] Thus, while Theophrastus did recognize that the effects of soil and climate affected plant growth, no conclusions regarding his espousal of a Lamarckian doctrine seem warranted from remarks such as these.

Before we leave Theophrastus, it is worth noting that while he may have been a devoted pupil of Aristotle's, he was not a slavish disciple. On the contrary, when he felt that he could not accept Aristotle's conclusions, he openly admitted his disagreement. This is especially apparent in his criticism of Aristotle's teleological principles. Theophrastus, in his usual cautionary manner, summarizes his position in the following words: "As regards the view that everything has a purpose and nothing is in vain, first of all the definition of this purpose is not so easy, as is often said; for where should we begin and where decide to stop? Moreover, it does not seem to be true of various things, some of which are due to chance and others to a certain necessity, as we see in the heavens and in many phenomena on earth."[5]

7
The Quest for Self-Awareness

When Theophrastus died (287 B.C.) there was no longer anyone prestigious enough to attract incipient scientists to the Lyceum, or to Athens itself, for that matter. Indeed, post-Aristotelian philosophy as a whole has been called "the story of decay,"[1] the mark of which was an "intense subjectivism" that began to permeate much of the Hellenistic world. Now the only knowledge considered worthy of investigation was that which helped one to determine what was becoming and what improper in one's conduct.

At the forefront of this quest for self-awareness were the two emergent schools of Stoicism and Epicureanism, both of which maintained that the only subjects of concern to man ought to be ethics and morality. Knowledge of the world was illusory; hence it was not worth pursuing. There was no room for any inquiry outside that dealing directly with man. Nevertheless, both schools still had something to say concerning the cosmological problem posed by pre-Socratic philosophy.

The Stoics contended that since all information was based upon what one learned from the senses, it followed that everything had to have a materialistic basis. This being so, the identification of the *arche* once again became an interesting problem. For their part, the Stoics fell back on Heraclitus's

THE QUEST FOR SELF-AWARENESS 67

principle of fire. All things were once fire; God, the moving and organizing principle, was fire incarnate. When it came time for the world to appear, God changed part of his fiery self into air, then water, and then earth. This was how the cosmos had come into being. Little was said, however, relating to the origins of life.

In contrast to the Stoics, the Epicureans refused to accept the existence of a deity in any form whatsoever. Consequently, they dismissed outright any teleological formulations such as those advanced by Anaxagoras or Aristotle. Their position was the mechanistic doctrine of the Atomists. The most thorough presentation of their views was given by the Roman poet, T. Lucretius Carus (99-55 B.C.) in a work called *The Nature of Things* (*De Rerum Natura*), which describes the formation of the world and its inhabitants without any recourse to superstition or the supernatural.

His starting point was that, initially, the world consisted of nothing more than a vast expanse of randomly moving atoms. Some of these atoms joined for a short period of time and then separated; others fused permanently. A product of this chance process was the formation of the earth. Then came the appearance of life: "the new earth then first put forth grass and bushes and next gave birth to the races of mortal creatures springing up in many numbers and in diverse fashions."[2]

Critical of the pre-Socratic hypothesis that life had emerged from the water, Lucretius flatly stated that "no living creatures can have dropped from heaven nor can those belonging to the land have come out of salt pools."[3] Instead, he argued that the earth had given rise to the living directly: "we may see in fact living worms spring out of stinking dung, when the soaked earth has gotten putrid after excessive rains. . . . And many living creatures even now spring out of the earth taking form by rains and the heat of the sun. . . . Mother Earth herself gave birth to mankind and at a time nearly fixed, shed forth every beast that ranges wildly over the great mountains, and at the same time, the fowls of the air with all their varied shapes." Thus, "the earth with good reason has gotten and keeps the title, Mother."[4]

But because she is in fact like a human mother, the earth "must have some limit set to her bearing." Consequently, the life-producing capacity of the earth ended when it "ceased like a woman worn out by length of days."[5] This then, was Lucretius's image of how life had come into being.

Since Lucretius adhered to the mechanistic view of Nature, he felt somewhat compelled to recognize Empedocles' point that "ill-formed" creatures may have inherited the earth at one time. Their existence, however, was short-lived, for "nature set a ban on their increase and they could not reach the coveted flower of age nor find food nor be united in marriage."[6]

Here then, is a foreshadowing of the Darwinian concept of natural selection. While many different species had spontaneously emerged from the earth, only some of these species were able to survive; the others perished. Those able to find food and suitable mates left offspring; those that did not, became extinct.

Lucretius also touched upon the possibility of the evolution of one species from another, but discounted it, for "if the first-beginnings of the things could be vanquished and changed, it would be uncertain too, what could and what could not rise into being . . . now could the generations reproduce so often each after its own kind the natural habits and way of life and motions of the parents." While species themselves could not develop into other species while they were alive, nonetheless, this might occur following death as a result of the atomic patterns of a given species coming apart at death and then the components being reformed into the bodies that were characteristic of other life forms: "Time changes the nature of the whole world and all things must pass from one condition to another and nothing continues like to itself: all things quit their bounds and all things nature changes and comes to alter."[7]

This formulation of Lucretius represents the most advanced development of the mechanistic account of Nature to be found in antiquity. Using the atom as his basic unit, Lucretius erected a concept of mother earth and placed within it the potential for the production of every living thing through the arrangement of its own atomic particles. Some of these creatures survived, some did not. Eventually, however, the generative capacity of the earth was exhausted and new forms of life ceased to be produced. This marked the end of the evolutionary process. Those life forms which had managed to survive the critical phase, continued to exist, being perpetuated through the normal means of propagation.

This approach might have had a great impact upon scientific thought had it not been for the influence of emergent Christianity, which looked upon it as the product of a group of heathens who denied the existence and influence of God. As a result, the theory lay in abeyance until well into the nineteenth century, when it was rediscovered and first applied to the science of chemistry by Dalton.

8
Early Christian Reappraisals

The early years of Christianity were a time of expectation. The end of the world was imminent. All thoughts were relegated to the Second Coming.

As time passed and the universe remained intact, some of the leading Christians also found time to think about their place in the continuing world and, not surprisingly, began to puzzle over the ageless question of man's beginnings.

Within the confines of their religion, however, the alternatives were limited. Mechanistic hypotheses like those proposed by the Atomists were unacceptable, of course. Lucretius and Epicurus were vilified as idiots by Church Fathers such as Lacantius (260–340).[1] The only acceptable answers were those which were based on the ancient writings of the Old Testament.

Yet, there were still those within the Church who recognized certain difficulties surrounding a literal interpretation of the Bible. For example, if the biblical account of the flood were true, the presence of animals in remote geographical areas defied explanation. This was the problem that challenged St. Augustine: "There is a question raised about all those kinds of beasts, which are not domesticated nor are produced from the earth like frogs, but are propagated by male and female parents, such as wolves and animals of that kind. It could be asked how they could be found in the islands after that flood in which all the animals not in the ark perished." Evidently some Christians still believed in spontaneous generation for the lower animals, but denied the possibility for animals at a more advanced phylogenic level. "It might indeed be said that they crossed to the islands by swimming, but this could only be true of those very near the mainland, whereas there are some so distant that it does not seem possible that any creature could reach them by swimming."[2] Unable to offer a rational explanation, Augustine turns to the irrational: "Some animals may have been captured by men and taken with them to those lands which they intended to inhabit, in order that they might have the pleasure of hunting; at the same time it cannot be denied that the transfer may have been accomplished through the intervention of angels, commanded or allowed to perform this labor by God."[3] Thus, while Augustine recognized that the distribution of various species presented a challenge to the Scriptures, he was still unable to contemplate a solution to this problem in terms that were outside the compass of theology.

A more basic problem that sorely perplexed many theologians, however, was the apparent phenomena of spontaneous generation itself. Genesis taught that God produced all the forms of life at the moment of Creation, but by 200 A.D., the list of animals still seen arising from the soil was seemingly endless. The following summary, given by Sextus Empiricus, a Greek physician living at Rome, describes these various creatures and the material from which they were seen to emerge: "as to origin, some animals are produced without sexual union, others by coition. And of those produced without coition, some come from fire, like the animalcules which appear in furnaces, others from putrid water, like gnats; others from wine when it turns sour, like ants; others from earth, like grasshoppers; others from marsh, like frogs; others from mud, like worms; others from asses, like beetles; others from greens, like caterpillars; others from fruits, like the gall-insects in wild figs; others from rotting animals, as bees from bulls and wasps from horses. Of the animals generated by coition, some—in fact the majority—come from homogeneous parents, others from heterogeneous parents, as do mules. Again, of animals in general, some are born alive, like men; others are born as eggs, like birds; and yet others as lumps of flesh, like bears. It is natural, then, that these dissimilar and variant modes of birth should produce much contrariety of sense-affection, and that this is a source of its divergent, discordant and conflicting character."[4]

Such apparently uncontrovertible evidence for the still-continuing creation of life from the soil forced in some Church Fathers a feeling of uneasiness, which resulted in their reappraising some of their theological beliefs.

For instance, St. Basil, the archbishop of Caesarea in Cappadocia, argued that the statement, "Let the earth bring forth living creatures," could be interpreted to mean that the order had continued so that the earth has not ceased to obey the Creator. "For if there are creatures which are successively produced by their predecessors, there are others that even today we see born from the earth itself."[5]

As Augustine put it, God "does not say, 'Let the seeds in the earth germinate the pasture grass and the fruitful tree,' but he says, 'Let the earth germinate the pasture grass sowing its seed.'" In other words, "the earth is said then to have produced grass and tree causally, that is, to have received the power of producing."[6] According to Augustine's interpretation, God had given the earth the potential for germination of plants and animals and this potential had never been withdrawn.

On the problem of spontaneous generation itself, Augustine commented, "a certain question also arises concerning some very tiny animals, namely, whether these were created in the first founding of things, or whether they result from the corruption of mortal things. For many of these arise either from the defects of living bodies, or from excrements, or exhalations, or from putrefaction of dead bodies, some also from the corrupting of trees and plants. Some from the corrupting of fruits. And we cannot rightly say that God is not the creator of all these. . . . It would be ridiculous to say that these were created when these animals themselves (i.e., those which appeared at the first creation) were produced, if it were not for the fact that there was already present in all animated bodies a certain natural force, as it were, preseminated, and as it were, the primordial beginnings of the future animals which were to arise according to the genera and differences of things, through the infallible administration of the unchangeable Creator who makes all things."[7] It may be noted that in hinting at the possibility that, in decaying, higher animals eventually give rise to various lower forms of life, Augustine is reiterating an idea previously advanced by Lucretius, a misguided pagan in the eyes of the Church.

But of all the insights and statements concerning the relationship between the various species of animals, by either the early Greek philosophers or those who followed, it is actually St. Basil who first recognized that two different species might somehow be related.

Beginning with the scriptural prologue, "Let the water bring forth abundantly many creatures that have life and fowl that may fly above the earth in the open firmament of heaven," St. Basil asked, "why do the waters give birth also to birds?" His answer, whether or not he recognized its meaning, was the first clear statement in antiquity of the theory of evolution. The waters also gave birth to birds "because there is, so to say, a family link between the creatures that fly and those that swim. Both are endowed with the property of swimming, their common derivation from the waters has made them of one family."[8]

In other words, a common ancestor had preceded these two different species of animals.

Unfortunately, these were the final statements of any merit on a subject that had challenged the curiosity of the Western world for a thousand years. It would take another thousand years before the Western world would again be free to rationally contemplate the origins of life.

9
Factors in the Decline of Science

By the second century A.D. scientific curiosity was virtually dead. Many historians feel that a major factor behind its demise was the loss of independence experienced by Greece after the time of Alexander the Great.[1] Since the domination of Greece, first by the Macedonians and then by the Romans, the Greek sense of self-pride had disintegrated leaving the people demoralized and insecure. Seeking a new meaning to life, the Greeks turned to the subjectivity of Stoicism and Epicureanism, philosophical schools that did not promote or even stimulate interests outside that dealing with the behavior of man.[2]

These philosophical movements soon passed from Greece to Rome, where they attained the status of religions and were consequently endowed with an aura of respectability and reverence. The effect of this assimilation into Roman culture meant that the kinds of questions that thoughtful Romans would be willing to consider would be those which had mainly to do with man's soul and little to do with his natural surroundings.

Another factor to be considered is the general attitude of the Romans themselves toward scientific inquiry. Although there was basically no obstacle preventing the Romans from exploring Nature, there was also no encouragement for them to do so. The Romans were essentially a practical people and as such they were interested in applicable results. Consequently, pure science had no meaning for them. Such activity was even discussed with contempt. Cicero, for example, actually lauded his fellow Romans because they were unlike the Greeks in such matters.[3] To receive approval and recognition, those who studied Nature had to demonstrate that by doing so there was something to be gained. Thus, physicians, anatomists, and botanists, such as Galen and Dioscorides, achieved fame, while natural philosophers such as Maximus of Tyre were, and still are, virtually unknown.[4]

The emergence of Christianity is often cited as a third factor in the decline of science. For example, Clodd[5] has argued that the early Christian belief that the Scriptures contained the true revelation of the origins of life, coupled with Christian disinterest in mundane problems, led to a general discouragement in scientific inquiry. But by the second century A.D., Christianity was still too insignificant a movement to have influenced the social classes that were most likely to have been involved in scientific investigation.[6] Furthermore, the scientific movement had already been arrested long before the coming of Christianity. The effect of the new religion was simply to add to the apathy already surrounding scientific inquiry.

Although the institution of slavery is sometimes also cited as putting an end to scientific and technological progress,[7] this too proves to be an inadequate explanation, for, as Sambursky[8] notes, one has only to cite the technological advancements of the Egyptians to refute that particular viewpoint.

Whatever the reason, one thing is clear: there was not the same enthusiasm for life in the Christian era as there had been during the period when Greece was in her glory. Perhaps the loss of purpose engenders a loss in human curiosity.

Epilogue

The belief in spontaneous generation as one of the factors underlying the origin of certain species was to be one of the most durable errors in the history of science. The theologians of the Middle Ages were especially responsive to the idea that animals could be produced through the combined action of water, air, heat, and decaying material, and in some quarters their credulity was stretched to the point of fantasy. For example, in the eleventh century there began to appear descriptions of goose-trees—trees that gave rise to geese. This strange wonder was soon embellished by the English encyclopaedist, Alexander Neckam (1157–1172), who wrote that these birds arose from the resin of pine trees when they came in contact with sea salt.[1] This belief still had currency in the fifteenth century, for in some quarters ducks were served on Fridays because, coming from the sea, they were classified as "fish" and were therefore immune from the prohibition against the eating of meat on that day. It is recorded, however, that at one such meal, a disgruntled guest was overheard to remark that his "fish" tasted very much like duck.

The concept of spontaneous generation was not seriously challenged, in fact, until the seventeenth century, when the Italian physician, F. Redi (1626–1698) introduced the first experimental evidence concerning its validity. Redi himself was convinced that the belief had arisen through faulty observation, but he was keenly aware of the Church's endorsement of the doctrine and he knew of the difficulties into which Galileo had gotten himself on account of his opposition to Church doctrine. Redi decided to steer a safer course. He granted that the earth and the creatures that populated it had all been initially created by God, but he challenged the contention that creation was still going on. "Although it be a matter of daily observation that infinite numbers of worms are produced in dead bodies and decayed plants, I feel, I say, inclined to believe that these worms are all generated by insemination and that the putrified matter in which they are found has no other office than that of serving as a place, or suitable nest where animals deposit their eggs at the breeding season, and in which they also find nourishment; otherwise, I assert that nothing is ever generated therein."[3]

Unlike his predecessors, however, Redi was of the opinion that "belief would be in vain without the confirmation of experiment," and he prepared to put his opinion to the test. Into two vessels he placed various pieces of meat. One of the vessels he then securely covered with a piece of fine gauze; the other he left exposed to the air. The latter vessel was soon infested with white maggots. The vessel that had been covered was, by contrast, relatively free of infestation. The flies that had been attracted to the covered vessel by the odor of the decaying meat had only been able to lay their tiny eggs on the surface of the gauze. As a result, very few eggs had been able to fall into and therefore hatch in the meat itself.

Despite the import of these observations, the belief in the spontaneous generation of animal life continued to be supported by many of the leading biologists of the eighteenth century. It was only in the nineteenth century that the doctrine was finally dispelled by Louis Pasteur (1822–1895), who demonstrated that the air was alive with countless numbers of

EPILOGUE

organisms, such as those coming from the spores of plants or the ova of animalcules. Thus, even when the agents themselves could not be seen, their seeds and ova could still be carried by the air to various materials. The phenomenon of spontaneous generation of these creatures was thus due only to the fact that when they were airborne they were too small to be visible to the unaided eye.

Although Pasteur's impressive refutation of the doctrine of spontaneous generation occurred in 1860, a year after Darwin's momentous theory of evolution had actually appeared in print, Pasteur in no small way still contributed to the acceptance of Darwin's position. This was the case because the notion of the spontaneous generation of animal life as previously entertained was plainly anti-evolutionary in scope, since by intimating that every new species was created anew, it was also implying that species were fixed and therefore immutable. In addition, the doctrine was basically unscientific, for it was impossible to confirm. All that could be done was to disprove certain instances where previously it had been accepted. Thus, once the belief that animal life emerged spontaneously from the earth fell into disrepute, the only other scientific alternative was Darwin's theory of the origin of species.

But though Darwin's theory presents the most widely accepted account of how the various species came into existence, the problem of the origin of life itself is still to be solved. It is here that the doctrine of spontaneous generation was to reach fruition, for there are only two possibilities in answer to the problem. Either life was created *ex nihilo* by a supernatural power, or it arose from nonliving material. Of the two alternatives, it is the latter which is the more testable and therefore the more scientifically respectable. There are, in fact, two classic pieces of evidence to give credence to this second possibility. The first comes from an experiment conducted in 1928 by

Fredrich Wöhler. Working with a number of inorganic materials, Wöhler was able to produce the organic compound urea, thus demonstrating that this product of living organisms could be produced by nonliving material.

The second important piece of evidence appeared in 1953.[4] By passing an electric spark for a period of a week through a mixture of water vapor, methane, ammonia, and hydrogen, S. L. Miller was able to produce a number of complex amino acids, the basic material required to sustain life. Here was a definite indication that life could have arisen through the action of certain forces on some of the basic material thought to have been present at the very beginning of time.

Notes

NOTES TO CHAPTER 1

1. Aristotle *Metaphysics* 783.20, in Aristotle, *Works,* ed. W. D. Ross. 12 vols. (Oxford: Clarendon Press, 1928). Cf. Strabo *Geography* I.8, (New York: G. P. Putnam's Sons (Loeb Lib.), 1917): "The first historians and philosophers of Nature were writers of myths."
2. Cf. W. Windelband, *History of Ancient Philosophy* (New York: C. Scribner's Sons, 1899), p. 35.
3. Diodorus *History* I.10 (Loeb Lib.) (Cambridge, Mass.: Harvard University Press, 1962).
4. S. G. F. Brandon, *Creation Legends of the Ancient Near East* (London: Hodder and Stoughton, 1963), p. 61.
5. *Ibid.*
6. *Ancient Near Eastern Texts,* ed. J. B. Pritchard (Princeton, N.J.: Princeton University Press, 1950), p. 99b.
7. *Ibid.,* p. 68a.
8. *Genesis* 2:7. (Jerusalem Bible.)
9. *Ibid.,* 2:18.
10. Cf. T. H. Gaster, *Myth, Legend, and Customs in the Old Testament.* (New York: Harper and Row, 1969), p. 8ff.
11. *Works and Days* (Loeb Lib.) (New York: The Macmillan Co., 1926), pp. 60 ff.
12. Ovid *Metamorphoses* I.318 ff. (Loeb Lib.) (New York: G. P. Putnam's Sons, 1916).

NOTES TO CHAPTER 2

I

1. The means by which Greek chronologists fixed a date for a historical indi-

vidual was to take an outstanding event in his life as an indicant of his floruit or prime.
 2. Aristotle *Meta.* 983b 18.
 3. *De Iside et Osiride* 34 in G. S. Kirk and J. E. Raven, *The Presocratic Philosophers* (Cambridge: Cambridge University Press, 1957), no. 70. Henceforth, all citations from this work will be designated as K. and R.
 4. J. Burnet, *Early Greek Philosophy,* 4th ed. (London: A. and C. Black, 1930), p. 40.
 5. *Ibid.*
 6. E.g., J. Burnet, *Gr. Phil.* and T. Gomperz, *Greek Thinkers* (London: J. Murray Co., 1912).
 7. E.g., E. Zeller, *A History of Greek Philosophy* (London: Longmans Green & Co., 1881); C. H. Kahn, *Anaximander and the Origins of Greek Cosmology* (New York: Columbia University Press, 1960).
 8. The translations are those of J. H. Loenen, "Was Anaximander an Evolutionist?", *Mnemosyne* 7 (1954): 215-32.
 9. *Ibid.*
 10. K. and R., p. 142.
 11. Theophrastus *apud simplicium physics* 24.26 (K. and R., no. 143).
 12. Aetius 1.4.4 (K. and R., no. 163).
 13. Fragmenta 7.947F (K. and R., no. 146).
 14. As Kirk and Raven (p. 150) point out, even though Anaximenes based his conclusions on empirical evidence, his lack of thoroughness led him to an erroneous conclusion, since it is condensation that is related to heat and rarefaction to cold, rather than vice versa as Anaximenes had surmised.

II

 1. Diogenes Laertius *Vita philosophorum* 9.1 (K. and R., no. 193).
 2. Plato *Cratylus* 40 A (K. and R., no. 218).
 3. Fr. 1 (K. and R., no. 218).
 4. E. Nordenskiöld, *The History of Biology* (New York: A. A. Knopf, 1928), p. 14.
 5. Fr. 55 (K. and R., no. 200).

III

 1. Fr. 11 (K. and R., no. 169).
 2. Fr. 16 (K. and R., no. 171).
 3. Fr. 15 (K. and R., no. 172).
 4. Fr. 3 (K. and R., no. 173).
 5. Fr. 5,6 (K. and R., no. 174).
 6. Fr. 4 (K. and R., no. 175).
 7. Fr. 9 (K. and R., no. 184).
 8. Hippolytus *Refutatio omnium haeresium* 1.14.5 (K. and R., no. 187).

NOTES 83

IV

1. Fr. 7 (K. and R., no. 346).

V

1. Fr. 6 (K. and R., no. 417).
2. Aetius 5.19.5 (K. and R., no. 442).
3. Fr. 57 (K. and R., no. 443).
4. Fr. 58 (in K. Freeman, *Ancilla to the Pre-Socratic Philosophers*) (Cambridge, Mass.: Harvard University Press, 1948), p. 58.
5. Aetius 5.19.5 (K. and R., no. 442).
6. Fr. 59 (K. and R., no. 444).
7. Fr. 61 (K. and R., no. 446).
8. Aristotle *Physica* 88.198b.9 (K. and R., no. 447).
9. Fr. 6 (K. and R., no. 448).
10. Fr. 90 (Freeman, p. 61).
11. Fr. 97 (Freeman, p. 62).
12. Fr. 127 (Freeman, p. 66).
13. Fr. 108 (Freeman, p. 63).
14. Fr. 107 (Freeman, p. 63).
15. Simplicius *Physics* 371.33.
16. Aristotle *Physica* 198b.7.
17. P. 40.
18. *Ibid.*, p. 37.

VI

1. Plato *Apology* 6 D (K. and R., no. 493).
2. Lucretius *On the Nature of Things* (Indianapolis: Bobbs-Merrill, 1965), I.830.
3. Fr. 10 (Freeman, p. 84).
4. Fr. 15 (K. and R., no. 516).
5. Theophrastus *De historia plantarum* 3.1.4 (K. and R., no. 553).
6. Hippolytus *Ref.* 1.81 (K. and R., no. 53).

VII

1. Diogenes 9.44 quoted by S. Sambursky, *The Physical World of the Greeks* (London: Routledge and Paul, 1960), p. 107.
2. Aristotle *De Generatione et corruptione* 314a.
3. *Ibid.*, A8.325a (K. and R., no. 552).
4. Simplicius *de caelo* 242.1 (K. and R., no. 582).
5. Plutarch *Stromateis* 7, quoted by Sambursky, p. 109.
6. Hippolytus *Ref.* 1.13 (K. and R., no. 564).
7. Aristotle *Meta.* 1071b (quoted by Sambursky, p. 11).
8. Aristotle *de Caelo* 300b (quoted by Sambursky, p. 11).

9. Sambursky, pp. 115–16.
10. Democritus, fr. 11 (Freeman, p. 93).

VIII

1. Simplicius *Phys.* 25.1 (K. and R., no. 601).
2. Fr. 2 (K. and R., no. 602).
3. Fr. 5 (K. and R., no. 606).
4. Fr. 3 (K. and R., no. 604).
5. Fr. 4 (K. and R., no. 605).
6. Quoted by Burnet, pp. 354–55.
7. Diogenes Laertius II.17.
8. *Ibid.*
9. See W. K. C. Guthrie, *In the Beginning* (London: Methuen, 1957).
10. Diog. Laer. II.17.
11. *Ref.* 1.9 (quoted by Burnet, pp. 359–60).
12. E.g., Epicurus, fr. 333; Lucretius *Nature* 5.805 ff.
13. Dio Chrysostom *Orationes* 12.30.

NOTE TO CHAPTER 3

1. W. A. Heidel, *The Heroic Age of Science* (Baltimore, Md.: Williams and Wilkins, 1933), p. 39.

NOTES TO CHAPTER 5

1. Cf. B. A. G. Fuller, *History of Greek Philosophy* (New York: H. Holt Co., 1923), p. 98.
2. *Generatione animalium* 3.11.762.18.
3. *Historia animalium* 8.1.588b.4.
4. *H.A.* 5.1.539a.20 ff.
5. *H. A.* 6.15.569a.10 ff.
6. *G.A.* 3.11.754a.1 ff.
7. *G.A.* 2.3.736b.15 ff.
8. *Meta* 12.9.1074b.32 ff.
9. *H.A.* 8.1.588a.1 ff.
10. *Ibid.*
11. *G.A.* 2.1.73b.32 ff.
12. *Politica* 1269a.
13. *G.A.* 3.211.762b.

NOTES TO CHAPTER 6

1. *Enquiry into Plants* 2.1 ff.
2. *Ibid.*, 1.5.
3. E.g., E. J. Gardner, *History of Biology* (Minneapolis: Burgess, 1965), p. 39.
4. *Enquiry* 11.2.7.
5. *Meta.* 2.7.

NOTES TO CHAPTER 7

1. W. T. Stace, *A Critical History of Greek Philosophy* (London: Macmillan and Co., 1920), p. 339.
2. *Nature* 5.568 ff.
3. *Ibid.*
4. *Ibid.*
5. *Ibid.*, 5.780 ff.
6. *Ibid.*, 5.855 ff.
7. *Ibid.*, 5.780 ff.

NOTES TO CHAPTER 8

1. *Divine Institutes.*
2. *City of God* 16.7.
3. *Ibid.*
4. Pyrrhus I.40–44.
5. Quoted by E. C. Messenger, *Evolution and Theology* (New York: The Macmillan Co., 1932).
6. *De Gen. ad lit* 5.4.
7. *Ibid.*
8. *Homily* 8.

NOTES TO CHAPTER 9

1. E.g., E. Clodd, *Pioneers of Evolution from Thales to Huxley* (New York: D. Appleton & Co., 1897), p. 21; Nordenskiöld, p. 44; C. J. Singer, *A History of Biology* (New York: H. Schuman, 1950), p. 63.
2. E. V. Arnold, *Roman Stoicism* (Cambridge: Cambridge University Press, 1911), p. 4, however, suggests that philosophical movements such as Stoicism were the cause, not the effect, of the decline in the spirit of patriotism.

3. Cicero, *Tusculanes disputationes* 1.2; cf. A. Reymond, *History of the Sciences in Graeco-Roman Antiquity* (New York: Biblo and Tannen, 1963), p. 92.
4. Cf. M. J. Sirkes and C. Zirkle, *The Evolution of Biology* (New York: Ronald Press, 1964), pp. 60–61.
5. *Pioneers,* p. 51; cf. B. Farrington, *Science in Antiquity* (New York: Oxford University Press, 1969), p. 135.
6. Singer, p. 64.
7. Farrington, p. 149b.
8. *Physical World,* p. 226.

NOTES TO CHAPTER 10

1. A. I. Oparin, *Genesis and Evolutionary Development* (New York: Academic Press, 1968), p. 18.
2. *Ibid.*
3. Quoted by T. S. Hall, *A Source Book in Animal Biology* (New York: McGraw-Hill, 1964).
4. S. L. Miller, "A production of amino acids under possible premature earth conditions," *Science* 117 (1953): 528.

Bibliography

Aristotle. *Works.* Edited by W. D. Ross. 12 vols. Oxford, 1928.

Arnold, E. V. *Roman Stoicism.* Cambridge: Cambridge University Press, 1911.

Augustine, St. *City of God* (Loeb Library) Cambridge, Mass.: Harvard University Press, 1957.

Brandon, S. G. F. *Creation Legends of the Ancient Near East.* London: Hodder and Stoughton, 1963.

Bulloch, W. *The History of Bacteriology.* Oxford: Oxford University Press, 1960.

Burnet, J. *Early Greek Philosophy.* 4th ed. London: A. and C. Black, 1930.

Clagett, M. *Greek Science in Antiquity.* New York: Abelard-Schuman, 1955.

Cleve, F. M. *The Giants Of Pre-Socratic Greek Philosophy.* The Hague: M. Nijhoff, 1965.

Clodd, E. *Pioneers of Evolution From Thales to Huxley.* New York: D. Appleton and Company, 1897.

Cohen, M. R. and Drabkin, I. E. *A Source Book In Greek Science.* Cambridge, Mass.: Harvard University Press, 1948.

Cornford, F. M. *Before and after Socrates.* Cambridge: Cambridge University Press, 1932.

Dio Chrysostom. *Orations* (Loeb Library) New York: G. P. Putnam's Sons, 1932.

Diodorus Siculus. *History.* Cambridge, Mass.: Harvard University Press, 1960.

Fairbanks, A. *The First Philosophers Of Greece.* London: Paul, Trench, and Company, 1898.

Farrington, B. *Science in Antiquity.* New York: Oxford University Press, 1969.

Fothergill, P. G. *Historical Aspects of Organic Evolution.* London: Hollis and Carter, 1952.

Freeman, K. *Ancilla to the PreSocratic Philosophers.* Cambridge, Mass.: Harvard University Press, 1948.

———. *The Pre-Socratic Philosophers.* Oxford: Blackwell, 1949.

Fuller, B. A. G. *History of Greek Philosophy.* New York: H. Holt Company, 1923.

Gardner, E. J. *History of Biology.* 2nd ed. Minneapolis: Burgess, 1965.

Gaster, T. H. *Myth, Legend, and Custom in the Old Testament.* New York: Harper and Row, 1969.

Glass, B., ed. *Forerunners of Darwin* (1845–1859). Baltimore: Johns Hopkins Press, 1968.

Gomperz, T. *Greek Thinkers.* London: J. Murray Company, 1912.

Guthrie, W. K. C. *In The Beginning.* London: Methuen, 1957.

———. *A History of Greek Philosophy.* Cambridge: Cambridge University Press, 1962.

Hall, T. S. *A Source Book in Animal Biology.* New York: McGraw-Hill, 1964.

Heidel, W. A. *The Heroic Age of Science.* Baltimore: Williams and Wilkins, 1933.

Hesiod. *Works and Days.* (Loeb Library) New York: The Macmillan Company, 1926.

Jaeger, W. *The Theology of the Early Greek Philosophers.* Oxford: Clarendon Press, 1948.

Kahn, C. H. *Anaximander and the Origins of Greek Cosmology.* New York: Columbia University Press, 1960.

Kirk, G. S. and Raven, J. E. *The Presocratic Philosophers.* Cambridge: Cambridge University Press, 1957.

Lucretius. *On the Nature of Things.* Indianapolis: Bobbs-Merrill, 1965.

Loenen, J. K. "Was Anaximander an Evolutionist?" *Mnemosyne* 7 (1954): 215–32.

BIBLIOGRAPHY

Lones, T. E. *Aristotle's Researches In Natural Science.* London: West, Newman and Company, 1912.

Mason, S. F. *A History of the Sciences.* New York: Collier Books, 1962.
Masson, J. *The Atomic Theory of Lucretius.* London: G. Bell and Sons, 1884.
McClure, M. T. and Lattimore, R. *The Early Philosophers of Greece.* New York: D. Appleton-Century Company, 1935.
McLean, G. F. and Aspell, P. J. *Ancient Western Philosophy.* New York: D. Appleton-Century-Crofts, Inc., 1971.
Messenger, E. C. *Evolution and Theology.* New York: The Macmillan Company, 1932.
Miller, S. L. "A Production of Amino Acids Under Possible Primitive Earth Conditions." *Science* 117 (1953): 518-29.

Nordenskiöld, E. *The History of Biology.* New York: A. A. Knopf, 1928.

Oates, W. J. *The Stoic and Epicurean Philosophers.* New York: Random House, 1940.
Oparin, A. I. *Genesis and Evolutionary Development of Life.* New York: Academic Press, 1968.
Osborn, H. *From the Greeks to Darwin.* New York: The Macmillan Co., 1894.
Ovid. *Metamorphoses.* (Loeb Library) New York: G. P. Putnam's Sons, 1916.
Owens, J. *A History of Ancient Western Philosophy.* New York: Appleton-Century-Crofts, 1959.

Plato. The *"Timaeus."* New York: Harcourt, Brace and Company, 1937.
Pritchard, J. B., ed. *Ancient Near Eastern Texts Relating to the Old Testament.* Princeton, N.J.: Princeton University Press, 1950.

Regnéll, H. *Ancient Views on the Nature of Life.* Lund, Sweden: Gleerug, 1967.
Reymond, A. *History of the Sciences in Graeco-Roman Antiquity.* New York: Biblo and Tannen, 1963.
Robin, L. *Greek Thought and the Origins of the Scientific Spirit.* New York: A. A. Knopf, 1928.

Robinson, J. M. *An Introduction to Early Greek Philosophy.* Boston: Houghton Mifflin Company, 1968.

Sambursky, S. *The Physical World of the Greeks.* London: Routledge & Kegan Paul, Ltd., 1960.

Sarton, G. *A History of Science.* Cambridge, Mass.: Harvard University Press, 1952.

Scoon, R. *Greek Philosophy before Plato.* Princeton, N.J.: Princeton University Press, 1928.

Singer, C. J. *A History of Biology.* New York: H. Schuman, 1950.

Sirks, M. J. and Zirkle, C. *The Evolution of Biology.* New York: Ronald Press, 1964.

Stace, W. T. *A Critical History of Greek Philosophy.* London: Macmillan and Company, 1920.

Strabo. *Geography.* New York: G. P. Putnam's Sons, 1917.

Veitch, J. *Lucretius and the Atomic Theory.* Glasgow: J. Maclehose, 1875.

Wald, G. "The Origin of Life," *Scientific American* 191 (1954): 44–53.

Windelband, W. *History of Ancient Philosophy.* New York: Charles Scribner's Sons, 1899.

Zeller, E. *A History of Greek Philosophy.* New York: Longmans, Green and Company, 1881.

———. *Aristotle and the Earlier Peripatetics.* New York: Longmans, Green and Company, 1897.

———. *The Stoics, Epicureans, and Sceptics.* New York: Longmans, Green and Company, 1880.

Index

Abiogenesis: among Egyptians, 15–18; among Greeks, 16, 20–21; among Hebrews, 18–19; in Anaximander, 25–28; in Anaximenes, 28–30; in Anaxagoras, 41–43; in Aristotle, 61–64; in Augustine, 70–73; in Democritus, 45; in Diodorus, 16–17; in Diogenes, 47–48; in Heraclitus, 31–32; in Lucretius, 67–69; in Parmenides, 35–36; in Thales, 24; in Theophrastus, 65; in Xenophanes, 34

Acquired characteristics, their transmission: in Anaximander, 27; in Aristotle, 63; in Empedocles, 39; in Theophrastus, 65

Adaptation, problem of in organisms: in Anaxagoras, 41–42; in Anaximander, 27; in Democritus, 44–45; in Empedocles, 40–41; in relation to Intelligent Design, 42

Air: as a single basic substance, 28; as primary element, in Anaximander, 25; in Anaximenes, 28–30; in Diogenes, 47; in Empedocles, 37; can pass into other substances, 28

Analogy: between Earth and Motherhood in Archelaus, 49; between man and universe, in Anaxagoras, 42; in Lucretius, 68

Anaxagoras, 41–43, 46, 47, 51, 60, 67
Anaximander, 25, 30, 35, 51
Anaximenes, 28–30, 35, 47, 48, 51, 82n.

Apeiron, 25, 28, 31
Archelaus, 47–49
Aristotle: concept of evolution, 60–63; concept of primordial germs, 61; criticism of atomistic theory, 45–46; his natural philosophy, 60; on decay and permanence, 61; on Empedocles, 40; on Thales, 24; on the soul, 60–62; on transformation of substances, 61; relation to his predecessors, 59; teleological views, 60, 65

Assumptions about the universe: in Babylonia, 22; in Egypt, 22; in Greece, 23, 50; as impediment to growth of science, 22–23

Athens, 41, 54–55, 59, 66

Atoms: immutability, 43–44; indivisible, 44; infinite number, 44; motion, 44; shape, 44; in Democritus, 43–44; in Lucretius, 67–68

Atomic theory: against teleologism, 45; in origins of life, 44–45; in evolution, 45–46; See also Democritus, Leucippus, Lucretius

Augustine, St., 70–73

Babylonia: creation myths, 17–18; relation of climate to philosophy, 18; assumptions about universe, 22; scientific accomplishments, 22
Basil, St., 72–73
Bible, 70; creation stories, 18–20

Chance, 45–46

Change. *See* Mutability
Christianity: its influence on evolutionary doctrine, 70–74; disposition toward atomistic doctrine, 69
Cicero, 76–77
Conservation of matter, 43, 52

Darwin, 15, 25, 26, 39, 76
Democritus: his natural philosophy, 43; answer to Parmenides, 46; anticipation of materialism, 45; on origin of life, 45
Design: intelligent, in Anaxagoras, 41–42; in Aristotle, 60–61; by chance, in Democritus, 44–45; in Diogenes, 48; in Empedocles, 38
Diodorus of Sicily, 16–17
Diogenes Laertius: on Archelaus, 49; on Heraclitus, 30
Dioscorides, 76
Dio Chrysostom, 49

Earth: as basic substance, 31, 37; spontaneous generation from, in Archelaus, 49; in Basil, 72; in Dio Chrysostom, 49; in Empedocles, 37; in Lucretius, 67; in Sextus Empiricus, 71; in Theophrastus, 65
Empedocles: his natural philosophy, 37–38; pluralism, 37; concept of succession of life forms, 37–38; precursor of Darwinian theory, 39–40; struggle for survival, 40
Empty space, 35–36, 44–45
Epicurus, 45, 66, 67, 70, 75
Eros, 37–38, 52
Evolution: Darwinian, 15; precursors to theory, in Anaximenes, 29; in Aristotle, 15; in Basil, 73; in Empedocles, 37–40; in Heraclitus, 30–31; in Lucretius, 68; in Plato, 57–58; progressive evolution, in Empedocles, 39–40; stages in development of theory, 15

Fire, as primary element, in Anaximander, 25; in Anaximenes, 29; in

Archelaus, 48; in Empedocles, 37; in Heraclitus, 31, 51; in Stoic philosophy, 67
Fossils, as evidence of past history of the earth, 34

Galen, 76
Germs, preexistent: in Anaxagoras, 42–43, 51; in Aristotle, 61; in Augustine, 72
Gods, 15–16, 23, 25, 27, 33–34, 50; as Thought, 62
Greeks: belief in their being, autochthonous, 16; commerce with Egypt, 24; effect of war on their scientific spirit, 54–55, 75; influence of their religion on philosophy, 21, 50; shortcomings, 46–49; their creation myths, 20

Heraclitus, 30–31, 35, 51, 66
Hesiod, 20, 23–24, 33

Intelligence. *See* Nous

Lacantius, 70
Lamarck, 39, 65
Leucippus, 43, 51
Love. *See* Eros
Lucretius: his evolutionary views, 67–68; his natural philosophy, 67–68; his relation to atomists, 67; idea of natural selection, 68; on rearrangement of atoms to make new creatures, 68–69; on spontaneous generation, 67–68

Maximus of Tyre, 26
Middle Ages, 77
Miletus, 23–24, 30, 40
Miller, 80
Mind, 43. *See* Nous
Monism: in myth, 16; in Greek science, 24–25, 28–29, 42, 47, 51
Mutability, concept of: in Anaximander, 25–26; in Aristotle, 63; in Empedocles, 40–41; in Heraclitus,

INDEX

31; in Lucretius, 68-69; in Plato, 57
Myth: its character, 15; ubiquity of some types, 16; attacked by Xenophanes, 33

Nature: in Greek mythology, 15-16
Neckam, 77
Nous, 41-42, 48, 52

Osborn, 9, 40, 41

Parmenides, 34-35, 42, 43, 56
Pasteur, 78, 79
Philosophy: in relation to myth, 15, 81n
Plato: his concept of evolution, 58-59; his influence on Science, 56-58; his natural philosophy, 56; his teleology, 57
Pluralism: in Empedocles, 37; in Parmenides, 34-35, 42, 46; in Plato, 57
Primordial milky fluid, 49

Redi, 77, 78
Rome, 71, 75
Romans: their attitude to philosophy and science, 75; their character, 75-76

Seeds. See Germs

Sextus Empiricus, 71
Simplicius, 40
Slavery, 76
Soul, 57, 60-61
Spontaneous generation. See Abiogenesis
Stoics, 67, 75, 85n.
Strabo, 81n.
Strife, 37-38, 52

Teleology, 40-41, 65, 67
Temperature, as it affects evolution: in Archelaus, 48-49; in Diogenes, 47-48
Thales: choice of water as basic substance, 23-24; reasons for choice, 24-25; his place in science, 23, 25
Theophrastus: his natural philosophy, 64-65; criticism of Aristotle, 65; emphasis on observation, 64; his skepticism, 64-65; on teleology, 65

Water: as primary substance, in Anaximander, 25; in Anaximenes, 29; in Archelaus, 48; in Empedocles, 37; in Heraclitus, 31; in Lucretius, 67; in Thales, 24, 51; in modern evolutionary theory, 52
Wöhler, 80

Xenophanes, 32-33